Born This Way

Becoming, Being, and Understanding Scientists

Part 1: The Origins of Modern Biological Science

Richard A Lockshin, Ph.D.

Published by Richard A. Lockshin

Smashwords Edition

Copyright 2013 Richard A. Lockshin

Discover other titles by Richard A. Lockshin at Smashwords.com (coming soon: *Born This Way: How Science is Done and Practiced,* by Richard A. Lockshin; other titles through Amazon).

Cover Photo courtesy of Igor Siwanowicz, http://photo.net/photodb/user?user_id=1783374, reprinted with permission

Smashwords Edition, License Notes

This ebook is licensed for your personal enjoyment only. This ebook may not be re-sold or given away to other people. If you would like to share this book with another person, please purchase an additional copy for each recipient. If you're reading this book and did not purchase it, or it was not purchased for your use only, then please return to Smashwords.com and purchase your own copy. Thank you for respecting the hard work of this author.

TABLE OF CONTENTS

Preface 6
Chapter 1: The theory of evolution 12
Chapter 2: The voyage of the Beagle 23
Chapter 3: A question of time 32
Chapter 4: The age of the earth 35
Chapter 5: A question of mechanism--the driving forces of evolution 66
Chapter 6: Natural selection 95
Chapter 7: The rediscovery of Mendel 102
Chapter 8: The history of our planet 111
Chapter 9: Ages of earth 134
Chapter 10: From whence new species? 151
Chapter 11: What is a human? 165
Chapter 12: When did humans acquire a soul? 194
Chapter 13: The impact of evolutionary theory 199
About the author 224
End Notes 226

Born This Way

Becoming, Being, and Understanding Scientists Part 1

The Origins of Modern Biological Science

Richard A Lockshin

Preface

All children are born biologists. Watch any child between the ages of three and six. They are fascinated with worms, bugs, frogs, and in fact anything that moves. They also need to classify ("Is it a boy or a girl?") and understand what they can about it. Some of us never outgrow that childhood curiosity and fascination, and we become the biologists of this world. Why we persist in this stunted stage of development—not outgrowing our curiosity—is a matter for developmental psychologists, but a surprising number of practicing adult scientists had, as children, a nickname something like "Questions". Sometimes our parents help, tolerating and even encouraging the incessant questions and vermin brought into the house, as opposed to the parent who, fatigued and exasperated, tells a child "Stop bothering me with your questions." But in the end, we survive, asking throughout our lives what things are, how they work, why they are the way they are and not something else, and how they came to be.

Sometimes these questions lead to other questions: Why others don't ask questions; why people argue so adamantly against the theory of evolution, when to deny it means that one has to deny most other parts of science, which they readily accept; and why we keep trying to explain what we do, only to be told that we are geeks, the material is boring, dull, incomprehensible, or yucky, and that "normal" people are not science junkies.

Therein lies the origin of this book. On a faculty committee reviewing our Core Curriculum offerings, I described my frustration with our typical "Biology for Non-Scientists" offerings: A watered down version of the standard textbook, raced through in a large class, with multiple-choice tests based on isolated facts. "What was in bold-face type on page 432?" In other words, "Who cares whether you can differentiate between the Ordovician and Devonian?" was the way I characterized the tests: a survey of facts unconnected to ideas, the flow of history through science, the exploration of ideas, the questions that were asked and the ingenious ways people invented to answer them: the heart of what science was. No wonder students were bored with what we offered. I wanted to teach a course about the ideas, the questions, the quest. Not only would it be more interesting, it would better prepare students for the developments that were to come in their lifetimes.

So how could we do this? We had to talk about a question of sufficient scope to encompass all sorts of ideas. Although my own research was in developmental biology, the history of the recognition of evolution (or, better, natural selection) was more encompassing. Not only was it the most important idea of the 19th and probably 20th Centuries, it was, in social terms at least, a subject of controversy, something that could always catch the attention of students. One could explore successively why, surprisingly, evolution was not recognized earlier in history, what information gradually accumulated to make its recognition inevitable, how ideas were broached, challenged, tested, and confirmed or refuted, the extent to which social attitudes helped or impeded understanding and sometimes tragically distorted the meanings of discoveries, where we were going, and what lay in the future. Such a subject could reach from self-evident observations such as the similarities and differences among animals to the most abstruse and abstract considerations of molecular biology, and it could provide a forum in which students could evaluate the consequences of current trends, including destruction of habitat and extinction of species to global warming. In short, this would be a course that would teach science as scientists saw it, a realm of question and meaning, and a subject of importance to our species and our world.

That of course was a wildly optimistic assumption. Some students of course were confused by the approach, preferring the (to me) much more boring but predictable dreary procession of names and numbers, a necessary and not-expected-to-be-fulfilling drudge in which, given appropriate direction, one could memorize a sufficient number of bizarre terms to pass the course and, none-the-worse, get on with life. The standard offering was a course to which students were sentenced, but my version was also a difficult course to replicate. Many otherwise commendable teachers did not have the experience to draw analogies between 16th C poetry and 16th C scientific discoveries or between the technology of exploration and the development of evolutionary theory, or were uncomfortable straying outside the boundaries defined for the discipline of basic biology. On the other hand, some students were inspired by this approach, and their response encouraged me that it was possible to proselytize for the scientific method and the scientific approach to life. The course became a book, which I and others used.[i] I therefore offer this version as a less didactic, more open version of what it is like to be a scientist, with the stories and anecdotes that led, eventually, to our recognition that we are of this earth, with all the aspirations, limitations, and potential hazards of any species. In a second book[ii] I hope, likewise with anecdotes and stories, to explain what the rules of science are, including hypotheses, evidence, and control experiments. Maybe after reading these comments, you will look with a closer or more jaundiced eye at television commercials or feel more comfortable with evaluating the almost daily and often conflicting new claims regarding diets, the value of certain medicines, or exercise or lifestyle recommendations. Perhaps you can add your voice with more confidence to discussions about the warming of the earth, the role of humans in that warming, or the relative merits of hybrid cars, nuclear vs wind or solar energy, and the other myriad questions of the day. This is the responsibility of an informed citizenry, with "informed" meaning having the information AND being in a position to evaluate it. I hope so.

There is another element here. Science is fun and my intention is more to explain why we are scientists. To that end I would rather play the role of raconteur, and run through anecdotes describing how science works, the stories of science. Science is all about puzzles and games and mysteries. In my classes I often used clipart pictures of a detective to emphasize that the hunt for an explanation of how something worked played out exactly like a good mystery, the gradual accumulation of clues, the ingenious experiment (read: the trap that the detective lays for the unsuspecting miscreant), the ruling out of alternate solutions, and the final proof of the hypothesis. It is not nearly the "search for Truth" or "my science is pure and holy" style that so often crops up in television mysteries; it's the fun of being the child called to the stage in a magic show, able to watch the magician up close and having the chance to figure out how the trick is done, that drives us. Normally we identify questions that we think have value to mankind, and we set out to find how the system works. Perhaps it is an ancient instinct, derived from our earliest hunter-gatherer ancestors, and reflected in gamblers today. To an early hunter, be it animal or human, there is an advantage to understanding the system: "If the antelope came to the spring last evening, it is likely to come this evening." If you talk to anyone who plays the lottery or gambles, it is striking how many of them invest faith in "systems" that make little mathematical sense: "I chose every third odd number, and I was only one number from winning!" This too, however illogical, is an effort to "understand the system." Scientists too invest considerable emotional energy in trying to find "how the system works". The excitement of the hunt makes every new day the most exciting day to be alive and to practice science.

***Note on what you can do with this book: Electronic format has a size limit that does not permit large or detailed pictures. However, this type of science is often best explained by visual techniques, pictures or videos. Therefore most of the pictures are indicated by outlines or sketches and are hotlinked to the source. Some are to web pages, in which case the link is to the page and for some a mirror copy is available on a Flickr® site available for this book. Photos that are taken from the personal collection of Richard A Lockshin and Zahra Zakeri Lockshin are also available via hotlink on the Flickr® site,
http://www.flickr.com/photos/ral_bornthisway

I am very grateful to Igor Siwanowicz for permission to use his photographs as a cover for the book and as part of Figures 5.5 and 10.2. I must acknowledge the tremendous tolerance of my wife, Zahra Zakeri, an accomplished scientist in her own right, who has borne with my musings, flights of fancy, and digressions over many years, as well as taking usually much better quality pictures than mine, many of which grace this book; my children, Miriam and Nora, who not only survived the meanderings of my mind but often outreach me in their curiosity; and above all, my late parents, Samuel and Florence Levin Lockshin, who were bemused by my and my brother's explorations, often catching escaped frogs and mantises, or tolerating the various odors of assorted caged animals. I was truly astonished when, after we had caught some tadpoles and were draining the dirty pond water into the sink at a friend's house, my friend's father came home and was furious about what we were doing. Many students over the years also caught that bug. I thank everyone for that tolerance. Curiosity is a wonderful thing, and it keeps you from getting bored in school.

~~~~~

PART ONE: STORIES FROM EVOLUTION

"In Biology, nothing makes sense except in the light of evolution"—
Niko Tinbergen

~~~~~

Chapter 1: The Theory of Evolution

The view from 1800

We can begin to look at science by looking at the theory of evolution. Although until recently it was not an experimental science, there are several reasons to consider it a paradigm. The theory of evolution is, arguably, the most important idea of the 19th and perhaps 20th Centuries. It has led directly to some of the most dramatic advances in 20th and 21st C biomedical sciences, including genetics and the understanding of DNA, as well as, unfortunately, some of the worst social moves of all time. The language by which evolution is described, at least in its earlier forms, is neither abstruse nor hermetic. And the story of why such an important idea developed in the mid-19th C, neither earlier nor later, describes quite clearly the interaction that always exists between science and the conceptions and understanding of the society in which scientists work.

Two corrections right off the bat: First, we call it a theory not to suggest that it is a random guess, in the sense of "I have a theory as to why the team lost the game," but to mean, in the scientific sense, that there is an infinitesimal but not zero chance that a better hypothesis will come along to replace or substantially modify it. As Einstein said, "No amount of experimentation can ever prove me right; a single experiment can prove me wrong". Second, it is not the "theory of evolution" for evolution is a series of data indicating the change in form and types of creatures throughout the history of the earth. Darwin's theory is more properly called the theory of natural selection or descent with modification—the mechanism that is purported to explain the evolution that has occurred.

There are many issues to address: evidence that living things have changed in the history of the earth; the extent to which people appreciated these changes in early times; the reason why the subject became an important issue in the 19th century; how Darwin's hypothesis explains the data; what is not explained by Darwin's hypothesis; and why this idea unlike many others is considered controversial.

It is not at all obvious that the world changes. Suppose that you have no ability to travel, to read, or to receive any radio or television communications about any other country–in short, suppose that you are living as an individual in any time prior to the 15th century. A river might flood and cut a new channel, there might be a landslide on the mountain, or the winter might be warmer than you ever remembered it. Nevertheless, from your childhood until your old age, the river is still there, the mountain is still there, and someone always will remember a warmer or colder winter. This city may expand or shrink and people and wars will come and go. Overall, the world has not seriously changed. You have no reason to assume that the mountain or river will disappear, or that the land could ever be covered with ice. Now let's take a look at the animals and plants around us. For simplicity's sake, let's just consider birds. I live in a suburban community in the northeast of the United States. I can expect to see on my lawn the following birds: robins, cardinals, mourning doves, starlings, various sparrows, blue jays, chickadees, and finches. Migrating through, I might see Baltimore orioles, Canada geese, an occasional duck, red-shouldered and sparrow hawks, and, circling overhead, an osprey. Closer to the seaside...but you get the idea. The point is that I can identify each of these birds and that I will typically not confuse one species with another: the species are distinct and fixed. I will not confuse a dove with a robin, or mistake a pigeon in the city, even though pigeons come in many colors. The Greeks described this certainty as an idea (in the original sense that we now use the word "ideal"—a perfect form of, say, pigeon-ness, to which each individual pigeon aspired. Some would come closer than others.

Even extinction is likely to pass unnoticed. Consider a species of fish sought after and netted beyond its capacity to reproduce. It becomes more and more rare, gradually becoming so rare that fishermen expect to catch it only very occasionally. Finally, it is forgotten. The last people to have seen it assume that one day they will see it again, except that they never do. Finally, a new generation of fishermen arises, and they have never seen the fish, and this continues until it is finally forgotten. Species do not go from massive numbers to extinct within one person's memory, at least at a time when hunting is not terribly efficient and books do not typically describe rare species.

Thus, to our hypothetical citizen in pre-15th Century mode, there is no reason to assume that either the physical world or the biological world changes. To the monotheistic world of Judaism, Christianity, and Islam, it is perfectly reasonable to assume that all creatures existed on earth since creation and that they will persist until eternity. Some creatures do not live in my land. I have heard of lions, and have even seen reasonably accurate depictions of lions, since the Romans brought them from south and east of the Mediterranean, but I have no more reason to believe in lions than I do to believe in mantigores, griffins, basilisks, or dragons. If lions are not seen in Europe, it is because they choose not to come. They could easily walk to Europe around the eastern end of the Mediterranean.

Figure 1.1. From left to right, top to bottom: Basilisk[iii], Mantigore[iv], Dragon[v], Griffin[vi].

Thus there is no need for a concept that the world is different from what it once was or that it changes. Artists typically illustrate Biblical figures in the garments of the day, and with the physical characteristics of the artists' people. This neatly structured world began to disintegrate in the 15th C owing to several causes. Books began to proliferate, providing a basis on which to extend knowledge in time and in location. It was becoming possible to think about whether Romans, Greeks or, for that matter, the Arab or Persian world understood the world in the same manner that Europeans understood it. A growing technology, involving boats, materiel of warfare, and large buildings, necessitated an efficient search for minerals, raw materials, fuels, and—so that royalty could demonstrate their wealth—jewels. To support this growth, explorers had to understand the world more thoroughly, to predict where resources might be found. And, finally, explorers began to bring back samples and tales of far-off lands, including creatures never seen in Europe. All of these activities converged into a sense that the world was far more complex than Europeans to that point had understood.

European and other societies had some inkling of all the components of the theory of natural selection, but until the mid-19th C no one had assembled the components into a comprehensive theory. To do so required recognition, to the level of comfort, of both the ideas that the earth was very old and that species were not fixed and could change. Both of these ideas challenged the written Word (as translated from Hebrew through Aramaic through Greek through Latin into vernacular) and therefore acceptance of the possibility that the Bible, if accurate, was at the very least allegorical rather than literal. All of these realizations were gradually taking hold by the 19th C, so that by the time that Darwin wrote, he and other intellectuals were excited by the possibility of resolving a major question, "that great fact—that mystery of mysteries—the first appearance of new beings on this earth." But to do so required turning this curiosity into a true scientific question, that is, to formulate a hypothesis: a mechanism by which one phenomenon, natural selection, could explain another, descent with modification.

These words, rather simple and deceptively easy to comprehend by themselves, in combination imply complex and elaborate mechanisms: natural selection, overbreeding of all populations, competition among individuals for resources, survival of those best adapted to use the resources, and further breeding by only those better-adapted individuals who, passing their better characteristics onto their children, would by descent with modification, leave future generations better and better adapted to their station in life, thus changing the species. Could such a mechanism explain even the creation of new species? This was the breathtaking possibility faced by Darwin and Wallace. We write "Darwin and Wallace" because, contemporaneously with Darwin, Alfred Russel Wallace generated nearly the same hypothesis. In fact, a manuscript he sent to Darwin[vii] precipitated the effort by the friends of the scrupulous and obsessively thorough Darwin to force him to publish his ideas. Wallace was less well placed than Darwin and more diffident, and he did not follow through with his research into evolution as Darwin did. He is therefore less well known to the public, but the coincidence raises another hugely interesting question: why in science does more than one group make many great discoveries nearly simultaneously? We will encounter another such coincidence in the nearly simultaneous rediscovery, by three laboratories, of Mendel's work nearly 40 years after it was published. The answer is that the questions are always there—where do species come from?—but until the questions can be formulated as hypotheses they cannot be answered by science. To formulate hypotheses, one needs sufficient evidence, which must be seen with sufficient clarity to describe a cause; one needs to define a process or mechanism with sufficient precision that one can determine a result or effect that cannot be easily explained by any alternative mechanism; and one needs to define the effect with sufficient accuracy that the explanation is not drowned in ambiguity. To accomplish this goal, for the mechanism of evolution, we needed a profound knowledge of the age of the earth, of the biology of species, and of mechanisms of competition. These separate understandings coalesced in the mid-19th C, making it possible for Darwin, and for Wallace, to formulate a hypothesis concerning the origin of species. Since then, scientists have elaborated the hypothesis, documented its validity in the most stunning and unexpected ways, and occasionally polished the edges by amending aspects of the mechanisms, but by and large have confirmed the hypothesis to such an extent that we now describe it as a theory: a hypothesis so tested and confirmed by so many different types of experiments and observations that we would be truly astonished if we were to find a contradiction. We grace broad hypotheses with this title, such as the Theory of Relativity, which has been used to build nuclear reactors and bombs, to explain the heat of the sun, and to conceptually reach deep into the heavens. Any potential challenge to a theory is sufficiently astonishing to trigger an intense and sometimes newsworthy search to find what is wrong with the challenge. If a theory is so completely consistent that we would describe a violation of the theory a miracle—if a building were suddenly to rise into space— then we would upgrade the theory to the status of a law, as the Law of Gravity (which itself is a component of the Theory of Relativity). In fact, recently a group of researchers thought that they had detected particles moving faster than the speed of light, which should be impossible and therefore would invalidate the Theory of Relativity. They pleaded with the scientific community to search for an error in their methodology or calculations. Eventually, they detected a loose fiber optics cable that may have created the result[viii]. But first, what happened to make the mid-19th C so fruitful? We can call it the great intellectual syzygy, and look at the elements of understanding time, defining species, and recognizing mechanism.

The role of books:

The Gutenberg Bible was first published in approximately 1454. Probably fewer than 200 copies were actually printed, but the intellectual change was enormous. Previously scholars had to travel great distances to consult a few, laboriously hand-copied and potentially full of errors, copies of major treatises on the nature of the world (what we today would call science). These manuscripts were quite literally cloistered and not generally accessible. Once printing was commonly used, scholars across Europe could consult and discuss the same work, including both text and illustrations; more importantly, it became possible for a larger proportion of the population to consider, marvel at, and question what was written. Information was stable enough to be cogitated and even to be evaluated from the standpoints of logic and likelihood. Herbals and bestiaries, with their wonderful illustrations and sometimes quite colorful descriptions of behavior, could now be read by many, including scholars who lived in or traveled to lands where they might have expected to encounter some of these strange creatures. Public maps could depict the shorelines of known and just-encountered lands, providing sailors with information by which they could direct their ships. It set the stage for a more analytical approach to information. (Books were a remarkable facilitator but not a requirement. Although some Polynesian societies developed a now-lost form of writing, their sailors through oral tradition used a vast knowledge of the stars to sail with certainty from one island to another in the South Pacific.) Nevertheless, the expansion of knowledge made possible by printing was enormous, raising questions and forcing scholars to evaluate the validity of many presumed facts and the reliability of many authorities. Such challenges included the authority of the Roman scholar Celsus (Philippus Aureolus Theophrastus Bombastus von Hohenheim (16th C) renamed himself "Paracelsus" or "beyond Celsus"—but, in turn, we have the term "bombast" thanks to him.) and the great physician Galen. Vesalius (16th C) memorably challenged the authority of Galen, simultaneously insulting other professors of anatomy who still preached the teachings of Galen.

To get a slight taste of the enormity of this revolution, those of sufficient age might recall the state of knowledge and the style of scholarship before the electronic age.

When I wrote my doctoral thesis in the 1960's, there were no Xerox, no computers, no fax, and no Internet. One went to the library on a regular basis and pored through older books and recent journals. The available high-tech procedures consisted of huge volumes, updated monthly but not complete for at least a year, named *Chemical Abstracts* and *Biological Abstracts*, which sorted recent publications on the basis of keywords, displaying them in the context of surrounding words, so that one could scan the list looking for interesting articles. To submit a manuscript meant to prepare an original plus carbon copies, with original copies of the figures, and to mail it to the journal. Correspondence would take a few months, and the article, if accepted, might appear within the year. Travel to meetings by air was not common, and scientists attended mostly local meetings. If one wanted to explore a new idea, the only option was to hunt through books in the library for information, with the guidance of a skilled librarian or colleague who knew something of the field. One could not "Google" information or look it up in Wikipedia. If, for instance, one was curious to know the shape of a bat's sternum—believe it or not, there is a reason for this—the only way to find out would be first, to be lucky enough to live within reasonable distance of a large library, go to the anatomy section of the medical sciences section, and hope to find a book on comparative anatomy of mammals that, hopefully, would contain something on bats and then look to see if there was any description or picture, in the right orientation, that would resolve the question. This enterprise might take a day or more. In the digital world, one can find information within minutes.

The role of technology:

By medieval times, the value of the resources of the earth was well understood. Fire-controlled extraction of metals was known, and alchemists were searching for new ways to mix materials to produce potentially valuable new compounds, potentially even to transmute dross materials into gold. Powerful acids such as aqua regia could dissolve even stable, "noble" metals such as gold and silver. Various earths were found to contain minerals of interesting and valuable properties. Gunpowder had reached Europe by the 13th C, marking the beginning of the end of the fortress cities and forcing feudal societies to consider alternative means of protecting themselves. Elaborate calculations of celestial movements permitted better estimate of times for planting and harvesting, and for the assignment of major holidays related to lunar and solar calendars, such as Easter: by the 16th C holes (apertures) were made in the great cathedrals, and meridian lines embedded into the floor, so that the image of the sun would cross the meridian line at noon and its position would mark the solstices and equinoxes[ix] so that one could recalibrate the calendar and assign the proper day for observing a holiday. More visibly, they mounted elaborate astronomical clocks[x], complex sundials that, from the position of the sun, marked not only the hours but also the days and seasons. Though more subdued, the interest of the Church in calibrating the year reflected the elaborate astrological constructs of Stone Age Europe.

As populations expanded in both numbers and in space occupied, increasing contacts brought new problems and required new technologies. Although there was no knowledge of vectors of disease such as bacteria or viruses, it was apparent that new and frightening diseases could be acquired, such as plague, first brought from Istanbul to Genoa, and that these diseases could spread rapidly in urban environments. It would be necessary to find means of controlling and isolating these diseases. Venice, a port of contact between East and West, contributed two words reflecting understanding of the time: "malaria" or "bad air" to explain why this disease appeared only in swampy areas; and "quarantine," a mandatory 40-day period during which incoming boats had to wait before disembarking, to establish that the passengers were free of disease.

The power of technology was manifest in the exploits of sailors and travelers. Instruments were devised to assess the positions of the sun and stars, astrolabes and sextants, so that seafaring boats could establish where they were and plot a safe return home. The success of the technology for all these purposes also drove an increasing interest in movement. Comprehension of movement had an immediate practical effect, such as in calculating or estimating the trajectory of materials launched from catapults or understanding the movement of tides—boats were getting larger, and it was becoming critical to predict when the tide would be sufficiently high to permit a ship to safely enter harbor. More intriguingly, movement could be considered to be the primary characteristic of life. How did a person or animal that has just expired differ from his, her, or its state a minute before? By movement of the chest, pulse, or eyes, or any other movement. It seemed that if one could understand movement one might be grasping one of the secrets of life. Thus all forms of movement—wind, gravity, trajectories of pendulums and thrown objects, tides, celestial bodies—were subjects of interest and curiosity. But overall was a gradually changing attitude: understanding the physical world had great practical value, and observation, intellectual analysis, and even a primitive form of experimentation, could profit kings and societies. The value of a secular understanding of the world was returning to recognition in Europe.

The role of exploration:

Travel and exploration both drove scientific and technical development and profited from it. In the most obvious situation, one cannot hope to travel around the world unless one believes that the world is round, as was becoming increasingly obvious from observations of the heavens and efforts to explain the changing seasons and the positions of the stars. Farmers were deeply interested in knowing the earliest and latest dates that they could hope to grow crops, and the Church needed to understand the relationship of lunar cycles to solar cycles in order to calculate calendar-dependent dates such as Easter. To this day, Islam and Judaism rely heavily on detecting the time of the new moon, while Zoroastrians time the beginning of the New Year to the minute of when the sun first crosses the equator at noon. Thus the Julian calendar, dating from 45 BC, was replaced in the 16th C by the Gregorian calendar, which more accurately calculated the solar year. The approximately 11-minute error in the Julian calendar had resulted in an approximate 11-day error by the 16th C, a problem both for farmers choosing dates on which to plant crops and for assessing Easter. These various corrections to the calendar explain why Christmas is on December 25 rather than the solstice date of December 21 or 22 and why Eastern Orthodox calendars differ from western calendars. Of more importance, the issue was of sufficient interest that cathedrals often had elaborate constructions designed to pinpoint the position of the sun.

Understanding time, the sun's movements, and seasons also contributed enormously to the ability to explore. Sailors had learned to judge how far north or south they were by comparing the highest point the sun attained at noon to tables that had been constructed for the season, but judging how far east or west they were was more complex. This could be very dangerous. By the 15th C, sailors could travel far out to sea, and by the 16th C they were bringing back riches from other lands. They sailed on trade winds: generally westward near the equator and eastward farther north, avoiding the area of still air (the doldrums) in between. Not only could they get lost en route to home, if they miscalculated and arrived offshore of a hostile country, their boats could be plundered. Therefore there were government- sponsored contests to find means of calculating longitude (east-west position). Galileo, in response to such a contest, even proposed measuring the positions of the moons of Jupiter.

This science-and-technology driven exploration led to new and troubling questions. The reports of the returning explorers challenged common thinking about the nature of the biological world, while the need to understand the nature of the planet began to challenge the Biblical definition of the age of the earth[xi].

Chapter 2: **The voyage of the Beagle**[i]

This then was the world and understanding that faced Darwin as he approached the end of his education. He had been recognized as being very curious and observant, but was considered to otherwise be an indifferent student. Descending from a family that included Erasmus Darwin, an early theoretician of the structure of the biological world, he was expected to become a physician. However, his experience with surgery before the era of anesthesia and sterile technique dissuaded him from that path. He next aspired to enter the clergy, a respectable profession for a young well-to-do man, but he showed more enthusiasm for collecting beetles. When one of his professors told him that Captain Fitzroy was embarking on a 5-year trip to complete the exploration of areas of the Southern Hemisphere and that Fitzroy sought a suitably well-bred good conversationalist to accompany him as the ship's naturalist, young Charles sought his father's permission. His father Robert, despairing that the young man would never settle down into a fixed career, told him that, if he could secure the recommendation of one person whose judgment the father fully respected, he would have permission. Fortunately, Charles Darwin's uncle, Josiah Wedgewood (of Wedgewood china) considered—whether because he knew of Charles' intense curiosity and powers of observation, or because he thought that maybe five years at sea would cause him to reconsider his path in life, we do not know—that it would be a good idea, and Robert Darwin highly respected his brother-in-law.

Thus began a remarkable voyage. Though, somewhat churlishly, we have to concede that the initial voyages of unnamed humans out of Africa to populate the globe and Columbus' voyage to find a route to India must be counted, surely the exploratory voyage of the Beagle has to be counted as one of humankind's most remarkable trips.

Mechanism of atoll formation

On the voyage young Charles Darwin established himself as a keen, observant, and analytical naturalist, devising a comprehensible interpretation of the origin of atolls[ii] (rings of coral islands surrounding shallow lagoons, found in warm oceans) that has survived to this day. He communicated this by letter while the voyage continued, earning himself recognition before he returned to England. Briefly, he followed the terrain of what were often central pinnacles in atolls, including tracing their contours under the sea. He recognized that the beds of rivers on the central pinnacle often extended deep underwater, suggesting that the underwater part had once been above sea level. He further noted that live coral was restricted to very specific depths in the sea, its lower limit being determined by available light, since there is much photosynthesis in coral, and its upper limit by the agitation of waves, since the surface of coral is very fragile. He proposed that atolls started as volcanoes surrounded by rings of corals. The volcanoes gradually sank into the sea, with the coral rings continuing to grow upward to the surface of the sea as the volcano sank until finally there was no volcano, only a coral ring. See [animation](#)[iii].

Another note of social importance was that, in an age with no television, radio, cinema, or telephone, journals such as Darwin would write, describing exotic places, people, and things, were read with considerable eagerness and curiosity, much more so than we today would follow programs such as Discovery Channel or travel programs that show us exotic lands or palaces. Thus his writings would have high impact. On the trip Darwin made many observations of peculiar relationships in the biological world, relationships that would lead him to the greater pattern of life. There were three elements to what he needed to put it all together: the data about biological variation and the patterns of that variation; true comfort with a sense of the age of the earth; and a mechanism. He got the first two from this voyage, and the third from reading a book by Malthus shortly after his return. The fourth element, which became an issue in 1900, long after the first three were generally acknowledged, was the [mechanism of inheritance](#). Let's consider first the age of the earth.

He had taken as reading material for the voyage Lyell's Principles of Geology (what can we say? For scientists, the whole world is like a murder mystery or puzzle to work out, and factual, scholarly books are clues.) He had digested Lyell's ideas, recognizing that the great Pampas of Argentina appeared to be outwash from the Andes mountains:

> "My geological examination of the country generally created a good deal of surprise amongst the Chilenos: it was long before they could be convinced that I was not hunting for mines. This was sometimes troublesome: I found the most ready way of explaining my employment, was to ask them how it was that they themselves were not curious concerning earthquakes and volcanos?—why some springs were hot and others cold? – why there were mountains in Chile and not a hill in La Plata? These bare questions at once satisfied and silenced the greater number; some, however (like a few in England who are a century behindhand), thought that all such inquiries were useless and impious; and that it was quite sufficient that God had thus made the mountains."
> ([Voyage of the Beagle](#))[iv]

Earthquake at Concepción

But he had not truly internalized them until a remarkable excursion in Chile rudely brought the issue to his attention. During a layover of the ship in 1835, he had been climbing in the Andes, marveling that one could find fossil seashells as high as he got on the mountain and, he was told, even to 14,000 feet (note: at this altitude most people have serious altitude sickness) and that the higher he got, the less like the modern fossils they appeared. Lyell had described this, but Darwin was now contemplating whether it could be true that the Andes Cordillera, which rose to 23,000 feet, could have once been under water. During this excursion, there was a violent earthquake near Concepcíon, and when it had ceased, what formerly had been seabed was now lifted 1-3 feet above the high tide limit. Shellfish that previously had dwelt near the shore were dying on these shelves, as he notes in *Voyage of the Beagle*[v]:

> "The most remarkable effect of this earthquake was the permanent elevation of the land, it would probably be far more correct to speak of it as the cause. There can be no doubt that the land round the Bay of Concepcion was upraised two or three feet; but it deserves notice, that owing to the wave having obliterated the old lines of tidal action on the sloping sandy shores, I could discover no evidence of this fact, except in the united testimony of the inhabitants, that one little rocky shoal, now exposed, was formerly covered with water. At the island of S. Maria (about thirty miles distant) the elevation was greater; on one part, Captain Fitz Roy founds beds of putrid mussel-shells _still adhering to the rocks_, ten feet above high-water mark: the inhabitants had formerly dived at lower-water spring-tides for these shells. The elevation of this province is particularly interesting, from its having been the theatre of several other violent earthquakes, and from the vast numbers of sea-shells scattered over the land, up to a height of certainly 600, and I believe, of 1000 feet. At Valparaiso, as I have remarked, similar shells are found at the height of 1300 feet: it is hardly possible to doubt that this great elevation has been effected by successive small uprisings, such as that which accompanied or caused the earthquake of this year, and likewise by an insensibly slow rise, which is certainly in progress on some parts of this coast."

He did not do the calculation at this point but, once he returned to England he considered it and realized what it meant. For instance, since Spanish records have been kept, between 1570 and 2012 there have been four major earthquakes near Concepcíon, or approximately one every 110 years. Assuming an uplift of two feet for each earthquake (see Fig. 10 on this website[vi]), it would take 770,000 years to lift the shoreline 14,000 feet and 1,265,000 years to lift it 23,000 feet. This calculation of course does not account for wearing down the peak of the mountain from the top.

Figure 2.1. Uplift of the Chilean coastline following the earthquake of 1835. The first shelf visible was underwater before the earthquake.

These and other observations left Darwin comfortable with Lyell's argument that major features of the earth had appeared gradually over remarkably prolonged periods of time, and thus prepared to accept the hypothesis that the Galapagos Islands, which he was soon to visit, and the Pacific atolls, which he had not yet seen, could have arisen from volcanoes that progressively rose and sank. Intellectually, the issue of time was now secure. We will return to the question, but at a minimum, Darwin was convinced that the world was older than calculations from the Bible would make it.

Sloths and armadillos

The next question concerned the many biological curiosities that he encountered. Sloths[vii], slow- moving, tree-dwelling mammals, were found only in the New World where, Darwin also learned, one could find fossils of giant, ground-dwelling sloths. The same was true for armadillos. This peculiar, primitive mammal still retains a scaly skin and scuttles around South America and southern North America; it is not found elsewhere. As was the case for sloths, fossils of giant armadillos[viii] could be found in South America and southern North America but only there. Why, he wondered, should one find fossils of strange animals only in the same locations in which one found living forms of these strange animals?

Figure 2.2. Left: Fossil skeleton of a giant armadillo. Center: A modern sloth. Right: Giant ground sloth, recovered from La Brea Tar Pits, California. Some of these creatures were over 17 feet (5.6 m) in height.

The rheas of South America presented a similar problem. Rheas are large, flightless birds similar to, but distinct from, African ostriches and Australian emus. (The similarity of these birds and their existence each on a separate southern hemisphere continent became an issue of considerable concern in evolution and ultimately led to a most interesting interpretation as described below. At this point, though, his question was more specific: rheas are found only in South America. In fact, there are two species of rhea, one slightly smaller and with slightly different colors. Darwin was the one who realized that they were two separate species, and pointed out that the ranges of the two abutted but did not overlap. This association of ranges piqued his curiosity—why did each species keep to its own range?— but he was more interested in a larger question: if all species were uniquely created, and there were two species of rhea, why should both species be found only in South America? What constrained the good Lord from placing one in Asia or North America?

The Galapagos Islands

The peculiarities of the animals of the Galapagos Islands[ix] deserve extensive consideration on their own accord, but in terms of the global distribution of animals Darwin was struck by one particularly curious aspect. Although the birds of the Galapagos were clearly independent species and different from any species elsewhere on earth, they were finches and more specifically finches strongly similar to the finches of the west coast of Ecuador. In contrast, the birds of the Cabo Verde Islands[x] strongly resembled those of the west African coast. This did not make sense. The Cabo Verde Islands are approximately 200 miles west of Africa, while the Galapagos are a bit less than 700 miles from Ecuador. The two groups of islands have relatively similar geography and climates. Why should each group of islands have its own indigenous species, and why, if species were created independently, should each group contain only species that resembled those on a nearby continent? Why, for instance, should the Galapagos finches not be found on the Cabo Verde Islands?

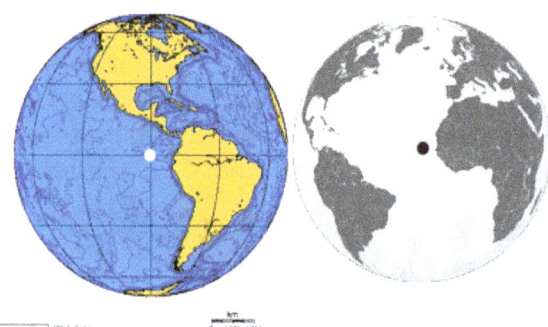

Figure 2.3. Position of the Galapagos Islands (left) and Cabo Verde Islands (right). Modified from sources by marking the positions of the islands.

A final anomaly that really caught his attention was the coconut-eating crab of Mauritius. This huge and remarkable creature is very well adapted to its peculiar existence. To open a coconut, it hammers at one of the eyes with its powerful pincer claw. Once it has made a crack, it uses its smaller hind claws to widen the hole and finally to reach inside and scoop out the flesh of the coconut. How, Darwin wondered, was it possible for a creature to come into being so perfectly adapted to feed off another species (the coconut)? Such interaction between two species (often plant and animal, but sometimes animal-animal or plant-plant) fascinated him: how could two species coordinate so well? He followed this issue for many such interactions. The most famous was his story of the star orchid, now known as Darwin's star orchid[xi]. This orchid from Madagascar has a spur or corolla up to 12 inches in length, at the bottom of which is found nectar to attract insects. Studying this structure in 1862, Darwin hypothesized that there must be an insect, which he presumed would be a hawkmoth, that would have a tongue or proboscis long enough to reach the nectar. *"It is, however, surprising that any insect should be able to reach the nectar: our English sphinxes have probosces as long as their bodies: but in Madagascar there must be moths with probosces capable of extension to a length of between ten and eleven inches!"* Some 40 years later, such a moth[xii] was found. In honor of Darwin's prediction, it was given the subspecific name *praedicta* (predicted).

Figure 2.4. Left: Coconut-eating crab. Center: Darwin's Star Orchid, corolla indicated by white arrow; Right: Darwin's Predicted Hawkmoth, tongue indicated by black arrow

Thus Darwin's readings and observations addressed the first big restriction—time—and his collections from the voyage led him to overcome the second great restriction. He now knew that animal species varied around the globe, and in a non-random manner. Many organisms appeared on one continent but not another, often accompanied by fossils of the same type; several species of finch, related to a South American finch, could exist on the Galapagos Islands, while birds related to African birds inhabited the Cabo Verde Islands; and many organisms were spectacularly adapted to another organism. He knew that these cues were suggesting that new species could arise on earth, rather than all species having been created at one time by a single act of Creation or having been collected on Noah's Ark. The question was how: if each creature was not a direct and specific (sorry) reflection of the Will of God, and if new life was not spontaneously generated from mud, by what mechanism did new life forms appear on earth?[xiii]

~~~~~

Chapter 3 A question of time

The Convergence of Data

One of the most powerful means of supporting an argument is the convergence of data from many independent sources. For instance, in a hearing about an automobile accident, Mary's statement, "Bill said that the blue car was going at least 70 miles per hour" does not independently confirm Bill's estimate and contribute as a second documentation of the speed of the blue car. However, any of the following would provide independent documentation of Bill's estimate: records from a traffic camera that recorded the car's speed; length of skid marks, from which one can calculate initial speed; amount of damage, from which engineers can estimate force of impact, which depends on speed; a bartender's statement that the driver left the bar just as the evening news came on (7:00 PM) and the accident occurred fifteen minutes later, twenty miles from the bar. Each of these is an independent measurement, based on different means of measurement, and thus corroborates the other estimate. If any of them conflict, a good lawyer can make hash of the charge of speeding. However, if all are consistent, then the presumption of speed is very high.

The same rules apply in science. Note that different, independent, methods of measurement are required. Two different measurements that both lead to a conclusion based on the mass or weight of a material may not be sufficient, as the balance may not be properly calibrated or there may be a systematic error that has not been accounted for. For instance, in certain circumstances a hot object may appear to be lighter than an object at room temperature, because the air rising from the hot object may lift the balance slightly. The apparent contradiction of the Theory of Relativity, [mentioned above](), is a case in point. In 2011-2012 an Italian group reported a measurement that appeared to show a subatomic particle traveling faster than the speed of light. They had shot such a particle from Switzerland to Italy, and it had arrived slightly earlier than expected, of the order of billionths of a second. Such a finding would indicate that Einstein's Theory of Relativity was at least partly wrong, a very difficult notion to swallow since so many predictions based on the theory had proven correct and would be otherwise unexplainable. Overthrowing the theory would have meant going back to the drawing board for many analyses and devices built on the basis of the theory, all of which worked. Therefore the team published a request to other scientists to help them find the error if it existed. Many possibilities were considered: that the team had not considered the relative movement of the satellite that recorded the time of travel between the launching and arrival; that there was an error in calibration of the clocks; that there were errors in the calculation; and many other sources of error. Finally, the problem was tracked to a weak electrical connection that very slightly delayed the processing of some critical information.

In the (unnecessary) controversy over "evolution" there remains some healthy if limited scientific argument over the basic mechanism, natural selection. Good scientists dispute the relative importance of random vs clearly selective events, as discussed in [Chapter 10](). This is normal and healthy, as science would not exist if no one continually challenged hypotheses. However, the public controversy focuses on two issues, both of which seem to be at odds with holy writings: the age of the earth and the relationship of humans to animals (and of animals to other animals and to all living things). It should be obvious that the fact that Judaism, Christianity, and Islam all describe essentially the same sequence of Creation does not constitute confirmation of the hypothesis, as they all derive from the same source, as was the case with Mary's account of Bill's comments. Unfortunately, this argument surfaces from time to time. For both of the fundamental issues, observation and theory in science contradict the story in Genesis (as interpreted by later scholars), but the evidence from multiple, independent sources is so overwhelming as to be incontrovertible. Let's look at each of these separately.

Chapter 4: The age of the Earth

The daring proposal broached in *Origin of the Species* depended on an earth sufficiently old to permit evolution to occur. The evidence for great age of the earth was to some extent available but not understood. There are many other ways to document the age of the earth, or to demonstrate at the very least that the earth was once very different from the way it looks today. Two of the most Eureka (Archimedes: "I found it!") moments were the recognition by Louis Agassiz of periods of immense glaciation (the "ice ages") and the recognition of continental drift. Another was the identification of a Jurassic-Era comet that may have contributed to the destruction of the dinosaurs. All of these are wonderful detective stories, the relentless and thoughtful pursuit of clues until the scientists concerned found the solution. If lay people looked at science as a series of murder mysteries rather than the hallmarks of geekdom, we would have a much better informed public. Detective stories are cool, and most of science can be looked at as a pursuit of marvelous detective stories. I often describe the feeling of working in science as feeling as if one is always ten years old and invited up to the stage to witness the magic trick and to see how it is done. Another description I could give is, "Every day is the most exciting day to be working in science," because every day brings new answers to ponder, and ideas to explore. Science is an onion. Each layer that one peels off reveals a new layer to explore underneath.

The wealth of nobility and royalty derives from many sources and is inevitably reflected in the acquisition by these leaders of the finest and rarest objects that they can acquire, whether it be feathers, skins, teeth, ivory, jewels, gold, or other fine and rare goods. Since the stone ages, warriors have required the accouterments of their craft, ranging from flint (which fractures leaving very sharp edges suitable for knives) to true, straight wood for arrows to steel, gunpowder and, today, radioactive earths. All building requires building materials and means to transport them, and households require utensils made of clay, stone, or steel. Fires for cooking, heating, and smelting of metals require wood and combustible materials such as oil and tar. All of these enterprises require exploration and knowledge of the resources. Even clay for utensils comes only from certain fine soils. Obviously one can scour the countryside looking for loose diamonds or flint, but those who can identify likely sources of such materials will come out ahead. Even stone-age peoples were known to dig ten or more feet into the ground to extricate flint. With increasing knowledge and books to retain that knowledge, descriptions of the earth began to appear. With these descriptions came new realizations and thus further questions. Some of the understanding came from relatively straightforward thinking and was known, for instance, to the Chinese by the 8th C, to the Islamic world (Avicenna and al-Biruni) by the 10th C, and to Native Americans in pre-Columbian times. It came later to Europe.

Genesis does not state an age for the Earth, but biblical scholars since at least Talmudic times have scoured the Holy Books for clues. Since much of the Old

Testament is a linear description of the history of the Hebrew peoples, the primary tactic was to sum the known information, making corrections and assumptions as warranted. For instance, the converted Jewish Roman scholar Josephus noted that according to the Gospels Herod reigned at Jesus' birth, but that Herod died in what would now be 4 BC, so that his birth could not have been at what we assign Year 1. Using these calculations, by the 17th C most scholars had identified a chronology beginning a few thousand years BC, when Bishop James Ussher, Primate of Ireland, attempted a more complete and definitive calculation. He used biblical indicators as a starting point, correcting for leap years and adjusting to known dates in early historical writings (including Egyptian, Persian and Babylonian dynasties and Greek and Roman chronologies). He also made a few assumptions as to the meanings of some ambiguous references and gaps in records. Among his other accomplishments before he died in 1656, he published, between 1650 and 1654, a major work delightfully entitled "Annals of the Old Testament, deduced from the first origins of the world, the chronicle of Asiatic and Egyptian matters together produced from the beginning of historical time up to the beginnings of Maccabes". He was following a respected scholarly tradition in trying to calculate the date and time of the origin of the earth. It was a prodigious undertaking. He read and compared ancient Hebrew, Greek, Roman, and Persian texts, calculated the dates of birth and death of prophets, compared dates to secular texts describing the lives of Herod, Nebuchadnezzar, and others; and consulted tables of calculations to determine the days of solstices and equinoxes as far back as necessary, and adjustments of the calendar. Ultimately, he determined that the Earth was created just before nightfall on October 23, 4004 BC. (The figure 4004 is interesting. Allowing for the error of four years, Ussher subscribed to millennial theory, in which great events occur at the millennia. Thus the Earth was created in 4004 BC, Solomon's First Temple was built in 1004 BC, Christ was born at 4 BC, and the Earth might end two thousand years later with the Second Coming of Christ, in 1996.) This is the calculation on which most western "young earth" hypotheses are built. Genesis itself is not nearly so explicit.

For over 200 years, this calculation remained (in Europe) the standard assumption of the age of the earth. James Hutton had begun promulgating, between 1785 and 1788, a theory that was subsequently called gradualism. By the late 18th C, scholars mostly agreed that the lines on the hills represented sedimentation, which they attributed to one or more catastrophic events such as Noah's Flood. Hutton suggested that catastrophes could not alone generate such patterns and that, rather, the sediments could have been built by the gradual accumulation of silt, which would have taken many years. He concluded, importantly, "The result, therefore, of our present enquiry is, that we find no vestige of a beginning,--no prospect of an end." It was a brilliant and provocative argument but, at over 2000 pages by the time it was completed in 1794, it understandably not widely read. Charles Lyell, by 1830, had been trained as a lawyer and therefore to make persuasive, digestible arguments. His interpretations of geological structures attracted attention and were widely read, so much so that he could support himself from his books alone. He was articulating the possibility that the earth was of great age. To Lyell, the processes that formed the earth would not differ from the processes observable today, as there was no justification for assuming that physical mechanisms such as gravity and friction had changed. Therefore, one could estimate the age of events by noting how they operated today. He was learning of the gigantic geological features of the New World. He measured rates of sedimentation from rivers and collected (somewhat inaccurate) information concerning the size of the Mississippi Delta. (Salt water can hold less sediment than fresh water, and the slowed rate of flow as the water disperses into the Gulf of Mexico also allows sediment to settle, so that mud accumulates and the delta gradually expands.) He also assumed that great falls such as (the by this time known) Niagara Falls could gradually chew their way back upriver into the higher lands from which the river came, and he could get an estimate of how rapidly such erosion occurred. From measurements such as these he tried to calculate how long it would take to form such features. His calculations led to values of tens or even hundreds of thousands of years—considerably longer than anyone had supposed. He dared to speculate that the Bible, and Bishop Ussher, were not accurate. Others throughout the 19th C found different means of extrapolating the age of the earth, coming to the same conclusions, but Darwin took Lyell's *Principles of Geology* with him when he sailed as a naturalist on the *Beagle*, and he greatly influenced Darwin. We describe below the types of evidence.

By the beginning of the 19th C many were increasingly convinced that the earth was much older than our predecessors had assumed, but by the 21st C we have much better means of calculating the age of the earth and, more important, several totally independent means of performing this calculation converge on the same figure. This principle, **independent means of verification**, is very important to science and to all logical thought. For instance, if witness Bill says that the car was going 90 miles/hour and witness Clara says that she heard from Bill that the car was going 90 miles/hour, Clara's statement does not constitute additional proof because it depends on Bill's estimate. If, however, if one can measure the length of skid marks (which will increase in length in some mathematical relationship to speed) or assess the damage caused by an accident (which will increase in relationship to speed), either of these will be additional proof, since they in no way depend on Bill's impression. Likewise, if we can find several means of measurement of the age of the earth that do not depend on each other, then each documentation becomes a verification of the other. To deny any single argument obliges one to find flaws with all the other arguments. Thus multiple independent means of verification is a powerful tool, and we can say with some confidence that the calculations that purport to be based on religious text are wrong. That is to say, there are four possibilities: Those who do the calculations based on religious text such as the Bible have made egregious mistakes; The Bible is wrong; The Bible is allegorical, not factual; or, contrary to the premise of rationality, the evidence of our senses, experimentation, and logic is not the most valid information in our universe. Part of the conflict between Creationists (at least young- earth creationists) and scientists is the rejection by each of the other's premise as to the most valid source of information in the universe.

Evidence for the age of the earth: Tree rings and local phenomena

Many other cultures in their creation myths ascribe an age to the earth of a few thousand years. Whether this consensus reflects a native human concept of "a very long time ago" or a reflection of the rise of consciousness (of the order of 50,000 years ago) or the appearance of civilization (10-5,000 years ago) cannot be determined. In contrast, observant humans have long noted indicators that the earth might be considerably more ancient. At a very minimum, once people realized that tree rings were formed annually (owing to faster growth in the spring than in the fall) it was a matter of patience to count the age of trees.

Some, like the bristlecone pines or the great redwoods of Muir Woods or the sequoias of Yosemite, proved to be thousands of years old. Better yet, by aligning rings from long-fallen trees, and even from petrified trees, one could extend the record to the range of 10,000 years and even read climatic conditions[i] through that period.

Figure 4.1. Left: Bristlecone pine[ii]. Center: Tree rings on fossil tree[iii] dated from 1245 to 1329. Right: Redwood in Muir Woods[iv], California that lived from 909 until 1930, 1021 years

Evidence for the age of the earth: Steno's laws

Nicolas Steno[v] was a remarkable man. A polymath, he contributed notable insights into biology as well as promulgating principles that became the foundations of geology, before finally being canonized in 1998, 302 years after his death. His insights into biology were related to his study of geology and could best be summarized as follows: On the island of Malta, some very curious formations known as tongue stones (glossopetrae) were found in the hills. These were attributed to miraculous origin, recollecting St. Peter's surviving a snakebite while on the island, and were sold (at considerable profit) as amulets. One day, however, some fishermen brought to Steno, who was at that time a renowned anatomist in Florence, a curious fish that they had caught. It was a great white shark, an open ocean, cooler water species that rarely enters the Mediterranean. Since these sharks grow to a size of up to 20 feet, Steno had the opportunity to examine it in detail and, in particular, its teeth. Most shark teeth are of the order of ½ inch to an inch or so in length.) In doing so, he realized that the teeth of the great white shark were remarkably similar to tongue stones, and he hypothesized that the <u>tongue stones</u>vi were in fact the teeth of truly giant sharks that had once existed but were no longer to be found in the seas. This led him to some further thoughts:

- The tongue stones could never have grown or been formed in the rock, because in that case their growth should have split the rock. Rather, they must have fallen into mud, which then congealed into rock.
- Mud can congeal into rock.
- Since sharks have no means to move about on land, the hills in which the tongue stones were found must at one time have been at the bottom of the sea, mud into which the teeth fell as sharks regularly shed their teeth.
- This must have happened long ago, at which time giant sharks were found in the sea; but such sharks no longer exist.

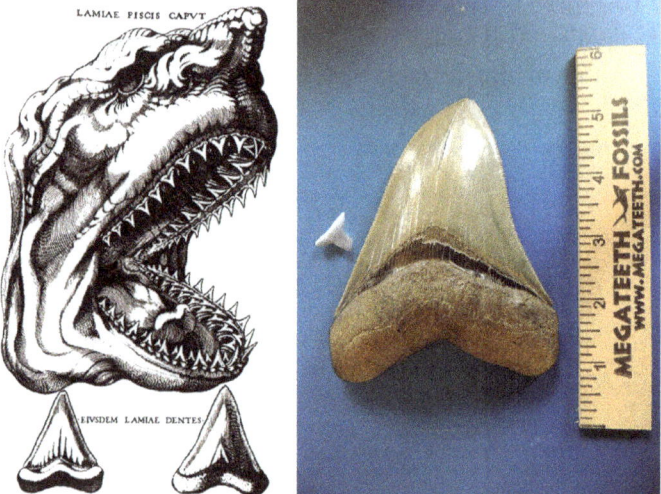

Figure 4.2. Steno's illustration that tongue stones were really shark teeth. Right: a megalodont tooth compared to that of a modern (approximately 6 ft) shark.

We have every reason today to believe that he was correct on all counts, and can date these fossils (see below) to 28-1.5 million years ago. We give the sharks that produced these teeth the name Megalodon (Greek for "giant tooth") and can calculate the size of the sharks as 50 to 67 feet in length. (A blue whale can reach 90 feet, but the larger whales typically feed on krill or small fish, while this was a true predator, with all the teeth and fearsomeness of today's sharks, but more so.) The reason that nobody had suspected the existence of such creatures is that sharks, unlike true vertebrates, do not have bony skeletons. Their skeletons, similar in shape to those of vertebrates, are composed entirely of cartilage, like our very immature forming bones and the joints between the bones. The teeth are calcified but are not true bone.

The stage was now set for admitting that the earth can change, but the issue of time was far from settled. Sedimentation patterns give no measurement of time. For instance, rivers entering the ocean deposit sediment, because moving water can suspend more sediment than relatively still water and because salt water can hold much less sediment than fresh water; once the two waters mix, the sediment will sink to the bottom. Suppose that you plant a yardstick into the sediment. You come back a year later and see that one inch of sediment has accumulated. Now you dig to bedrock and find that bedrock is four feet beneath the current layer of sediment. You can calculate that the sediment has been accumulating in this area for 48 years. If the sediment is interrupted by a layer of different color at, say, two feet, you might assume that sediment accumulated for 24 years, there was a different event, such as a flood or a drought, and then sediment accumulated for 24 more years. Likewise, if a river overflows its banks every spring and deposits on average 0.1" each time, a deposit of sediment two feet deep would reflect 240 years of floods.

Nothing, however, excludes the possibility that one massive flood could have deposited all or most of the sediment at one time. The different layers indicate different events and can even give the order of events (another of Steno's rules), but do not establish a true chronology. Even massive sedimentary fields—the Grand Canyon cuts through a sedimentary field over one mile deep—do not by themselves deny a chronology interpreted from the Bible of approximately 6000 years. Adapting to the concept of an ancient earth is necessary if one is to accept the idea that we descend from original, simpler creatures. Everyone could believe that species could be modified: dog breeders adapted breeds for specific purposes, such as shepherds, sheep dogs, pointers, badger hunters (dachshund), huskies and Great Pyrenees, even miniature lap dogs, pug-nosed dogs such as pugs and bulldogs, and racing dogs such as greyhounds. Plant breeders could select the seeds of the best plants or of the plants that had the most desirable characteristics and thereby larger, more colorful fruit, or change the length of the growing season or resistance to frost. These however were variations in a species, albeit sometimes dramatic, and not changes of species. Even if one were to dabble with the anti-Aristotelian view that species could change, the time required would be enormous. Selecting for a different breed of dog would take many years. If one imagined the conversion from wolf to dog, or from a dog-like creature (jackals, foxes, coyotes) or a long-ago ancestor that gave rise to both seals and dogs...one could even dare to consider that mammals and reptiles might be related, but to get this level of change would have to take much more than six thousand years. Thus there were two obstacles: accept that species can change, and find enough time for this to occur.

Tilting his hat more toward the geology side of his many capabilities, Steno noticed that the peculiar stripes and patterns found on the sides of hills on one side of a valley were mirrored in the patterns on the other side of the valley and indeed, if a white streak could be seen as a thin layer on one hillside, it could be found on another hillside miles away. He realized that the earth was composed of many layers, one on top of the other, that extended great distances. Looking further, he realized that some of these layers, though rocks, were remarkably similar to the muds that he knew to be deposited in slow-flowing rivers or their mouths: Flood waters would carry a considerable amount of debris. As the water slowed, a typical sediment would form, in which the larger pebbles would settle first, followed by the smaller and lighter ones. After the floods, the characteristics of the sediment would change a bit, depending on the situation. One layer would accumulate on top of the other. He surmised that the patterns reflected a single process across the entire region. Looking more closely, he applied simple logic to explain the patterns: many looked like the muds that rivers left behind in floods. If the stripes on the hills represented the detritus of massive floods, he could even interpret the pattern: When floodwaters settle, the heavier or denser stones sink first, with the lighter ones following.

Occasionally he could find rocks in which these lines—we will call them sedimentary lines—were tilted as substantial angles or even overturned. Even with these, he could still read the original positions by noting the order of large-to-small stones in each layer. This led him to further conclusions:

Figure 4.3. Uplifted and twisted sedimentation layers. Near San Andreas Fault, California.

- The different layers represented several periods during which the land was under water.
- These periods applied over huge areas and indicated several periods that could be described as flooding, as opposed to Noah's single flood.

- These layers could conceivably have been once at the bottom of the sea, in which case they had over an undetermined but lengthy period of time had risen above the sea.
- The land could tilt, as indicated by the sedimentation layers at angles. The tilting could produce hills and mountains.
- The land could erode again, returning to the sea.

Such conclusions had been reached much earlier by a somewhat secretive 10th C Islamic society named the Brethren of Purity who, building on the same foundation of recognizing the structure of layer patterns in the earth, described the building and erosion of the earth but, given their secretiveness and a European lack of interest in Islamic science, the idea had not attracted much attention in Europe.

Evidence for the age of the earth: Erosion

By the beginning of the 19th C, geologists were beginning to accept the idea that the processes observed day-to-day could explain, on much vaster scale, the features of the earth and, with greater relevance to us, could form a measure of age of the earth. If a river could gradually and measurably cut through earth to form steep banks, it could eventually form a canyon of substantial size. (The Grand Canyon had been seen by westerners long before, but it had not been considered in this context until the 19th C. However, there are major river-cut canyons in Europe that formed the basis of speculation.) Strong waves erode cliffs at the edge of the sea. . For instance, the Montauk Point Lighthouse guards the eastern tip of Long Island in New York state. It sits about 100 feet inland from a sharp bluff. However, when it was erected, it sat approximately 300 feet inland. Over the last 200 years the bluff has eroded at the rate of approximately one foot/year. This is longer than any one person's memory, and would perhaps not have been noticed if engineering diagrams from the early 19th C were not available.[vii] Many such relatively minor but nevertheless consequential changes have been noted in many parts of the world Since both still water and salt water can hold less silt than running, fresh water, the silt falls to the bottom as rivers flow into the sea, gradually expanding a delta where the river meets the sea.

Figure 4.4. Left: Sediment deposited at the mouth of the Mississippi[viii] River. Right: Erosion of cliff at Montauk, Long Island, New York, between 1838 and 1990. The cliff eroded 120 feet (approximately 40 meters) in 152 years.

Such changes can be measured over periods of a few years, a lifetime or, since there was now a history of accurate measurements, over periods of hundreds of years. These measurements indicate that, contrary to common sense, the land does change. Extrapolating the rate of change, as Lyell did for the retreat of Niagara Falls[ix] from the original escarpment back to the origin of the Niagara River at Lake Erie, or for the expansion of the Mississippi Delta,[x] one comes up with figures of minimally tens of thousands of years. The Grand Canyon is more impressive: It would have taken millions of years to cut its channel.

Figure 4.5. Niagara River. The falls started to cut at Lake Ontario (top, North) and has worked its way back to the point indicated by the arrow. Eventually it will cut back through to Lake Erie (bottom, South)

Although there are many reasons to argue that the extrapolations are not accurate, by the 19th C there were many reasons to assume that the Grand Canyon was an ancient formation. First, there are, unequivocally, fossil seashells at the top of the canyon (Fig. 11) and the surface at the top of the canyon is flat, like a seabed, even though it is 7,000 feet (2100 m) above sea level. Second, the current and past rates of cutting by the river can be calculated. Although there is variation, current estimates are of the order of 10 cm/1000 years or 0.01 cm/year (about 0.004"/yr). The canyon is 6,000 feet (1,800 m) deep. At the rate that the river is cutting and has cut between two measureable periods, if the rate of cutting were constant, it would have taken 18,000,000 years to cut the canyon. Obviously there are numerous sources of error—for instance, a strong flood might have cut much more in a single incident—but current estimates give a figure near to this, and they further indicate that up to 2 billion years of the earth's history is exposed in the Grand Canyon. No matter what Westerners thought and how they estimated the Grand Canyon during the 19th C, it was clear that erosion as understood then would have taken a huge amount of time, considerably more than Biblical time, to produce the canyon.

Figure 4.6. Fossils of marine worms and barnacles at the top of the Grand Canyon, over 6000 feet above sea level.

Evidence for the age of the earth: Glaciation

The 19th century brought some profound revisions to our concept of the history of the world. One of them was the realization that the earth had once been substantially covered with ice. Louis Agassiz, a Swiss paleontologist, was quite familiar with glaciers in the mountains of Switzerland. Glaciers are masses of ice that differ from ice packs in that they move. Owing to peculiarities in the physical chemistry of water, ice under pressure will melt. We see this in ice-skating, where the weight of one's body on the narrow surface of the blade causes the ice to melt underneath the blade and this allows us to glide easily along the ice. The weight the ice in a glacier causes the ice at the bottom to melt, and the glacier slowly slides down the mountain, moving a few feet to a mile or so per year. Of course, it is replenished by snowfall every winter, the snow packing to ice as the weight of more snow accumulates above it, and the glacier acts as a slowly moving river. This movement has consequences: The moving ice scours out the valley, leaving deep scratches on the bedrock and producing a characteristic long, narrow, steep-walled, U-shaped valley. The rubble accumulates along the sides and in front of the glacier, producing what are called, respectively, lateral and terminal moraines. Agassiz came to the United States, where he had been invited to join the faculty at Harvard, and he began to notice the landscape. Everywhere he looked, from the northern part of North America as far south as New York City, Ohio, or Colorado, he saw signs of glaciation. Rocks were deeply scratched, mostly in a north-south direction (Fig. 4.7).

Figure 4.7: Top left and center: Two examples of glacial striations, running North and South, in Central Park, New York City. Right: Shelter Rock, a glacial erratic, the largest of many found on Long Island, New York. Bottom: A smaller glacial erratic, also on Long Island.

There were steep valleys, like the fjords of Norway, in Maine, in the Hudson River Valley, and in the Finger Lakes district of New York. North of Long Island Sound, topsoil was very thin, with bedrock exposed or near the surface, and it was heavily scratched. Many of the hills, particularly those in a north-south direction, were masses of pebbles and rocks (lateral moraines). Long Island itself, an east-west island on its north shore was mostly hilly, with pebbles and sand, while the middle and southern parts of the island had much deeper soil, producing a quite different vegetation (Figure 4.8), very much like a terminal moraine. In various locations from Long Island northward were huge boulders that bore no resemblance to the surrounding soils. For instance, one on Long Island, truly massive in size, is similar to the granite found in New Hampshire and Vermont(Figure 4.7). Further west, the Great Lakes have the shape and characteristics, including pebbly soil to the north and deep topsoil to the south, that resemble the pools of water seen at the face of a land-locked glacier as it melts.

Figure 4.8. A wooded scene from southern Westchester County, New York (left), north of Long Island Sound and therefore of the moraine, and from Queens County, New York, on Long Island and on the moraine, slightly more

than 30 miles to the southeast. The two photographs were taken within two days of each other. Note the difference in vegetation. From shallow-rooted birches and aspens (left) to deeper-rooted oaks and maples.

All of this consolidated the idea that Agassiz had been developing, that there had been glaciers, far more massive than any that he had ever seen, that once covered North America as far south as Ohio. If this were true, then the characteristics of the land could tell him much about the glaciers. From the amount of damage, he realized that the ice in some places had been more than a mile deep. Erosion and sediment told him that the glaciers had existed very long ago. (If a river, for instance from melting ice, cuts through a geologic feature, obviously the feature existed before the river did. Judging the rate of erosion gives an estimate of how long the river has existed. A similar argument applies to sediment collected by action of wind or water.) The sense of the thrill of discovering the signs of the ice age can be captured by anyone living in northern North America. Once one is aware, one can find everywhere the signs that the land has at one point been very different. Another huge idea, which we can address later, was the one that is now known as [continental drift](), now well documented and quantified by many techniques.

Glaciation was another, not strictly relevant, but perturbing realization in the 19th C. It was perturbing because there was no mention of it in the Bible, and in any case if it had occurred it implied certainly longer stretches of time than Ussher's calculation.

Agassiz had begun to suspect that the steep-walled, long, narrow fjords of Norway were the remnants of much larger glaciers that had disappeared and that some of the other characteristics of northern Europe looked suspiciously glacial. Then he came to the New World. In the United States and Canada he found the unmistakable signs of truly vast glaciers everywhere he looked, throughout Canada and in the United States south about to Ohio: scoured rocks in New York's Central Park and around the Great Lakes; mountains such as Maine's Cadillac Mountain sloping far more gently on its northern than its southern side as if it had been ground down from the north; boulders sitting in the middle of soft terrain with the characteristics of stone mountains hundreds of miles away; lake structures resembling the terminal meltwaters at the front of a glacier. But note the size of the Great Lakes; smaller lakes in fjord-like valleys like the Finger Lakes of New York; long narrow aggregations of pebbles resembling lateral and terminal moraines, but hundreds of miles long (New York's Long Island to Cape Cod) and thick pebble- free soil south of all these marks but thin, pebble-filled soil north of them. (Glaciers scrape the topsoil away, piling it into moraines in front and to the sides of the glacier.) A few simple calculations gave rise to an image of vast ice sheets, several miles thick, covering the northern half of North America. You can imagine his sense of astonishment as the image of this vast, strange world began to take shape (Fig. 4.9).

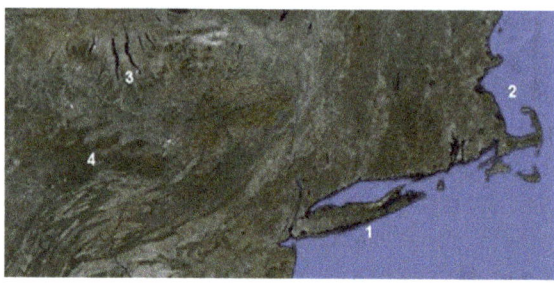

Figure 4.9: Upper: Near Cadillac Mountain, Maine[xi]: The glaciers came from the North (left) in the picture. Lower: Northeastern United States, illustrating the terminal moraine running from Long Island (1) to Cape Cod (2), The Finger Lakes (3) representing ice tongues from the glacier that ended in the Great Lakes to the north, and the folding the produced the Appalachian mountains (4) as a result of continental drift..

Evidence for the age of the earth: Mountain building

Reading something in a book makes you aware of it. Seeing it or experiencing it makes you think about it, seek to understand it, and appreciate its implications. As noted above, Darwin had taken Lyell's *Principles of Geology* along with him as reading material when he travelled on the Beagle. He mulled over the earthquake at Concepcìon quite a bit. By the time he was back in England, he began to appreciate that, by the crudest assumptions, simple mathematics would indicate that it would take 100,000 years to lift the mountains. This was a severe underestimate but it extended the age of the earth enough to begin to consider that there had been enough time for species to change.

Evidence for the age of the earth: Thermal cooling

Physics was a subject that attracted a lot of attention and heat, being the primary means of making machinery function as well as a characteristic of many living things, merited investigation. The laws of gravity and motion were consistent with the argument that planets had at one point pulled from the sun and thereafter orbited the sun. In this case, the planets would have once been hot and would have cooled, as smaller objects cool more quickly than larger objects[xii]. As William Thomson, Lord Kelvin and others had shown, the rate at which a hot object cools depends on its initial temperature, its thermal conductivity (the rate at which heat moves through an object: faster through metal than dirt, which is why you can put your hand on hot dirt but not metal at the same temperature), and it size. The size issue is straightforward. If you compare two balls, one of which has a radius 10x that of the other, the larger ball will have 100x the surface area (S = $4\pi r^2$) but 1000x the volume. Since the total heat it contains will depend on its volume, but it will lose the heat through its surface, the larger ball will cool at $1/10^{th}$ the rate of the smaller ball. Several other observations were consistent, such as the realization that, in mines, the earth got hotter the deeper you went into it (as a heated object cools more quickly on the surface than in the interior). One could even calculate the presumptive temperature at the center of the earth, and this calculation, too, was consistent with evidence from magnetism that the core of the earth might be fluid. Volcanoes would then be escapes of this inner molten core to the surface. Lord Kelvin had calculated the temperature of the sun, based on the knowledge that materials at different temperatures glow at different colors. Taking this information and the known laws of thermodynamics, he attempted to calculate, based on this hypothesis, how long the earth would have taken to cool to its present temperature. (Red hot steel is about 1800°K (2800° F), white hot steel is 5500° (9500° F), and a hot blue flame is 16,000° K (28,000° F)). He first calculated hundreds of millions of years. He was far from a supporter of the Darwin's ideas, and he redid his calculations several times, finally reducing the age to approximately 20 million years[xiii]. He was still very far off, because he was not aware of radioactivity. He published his paper in 1862k whereas radioactivity was discovered and first understood approximately 40 years later. Radioactive atoms, which occur naturally in the earth, continuously decay, releasing their energy, which ultimately appears as heat. This effect is sufficient to contribute substantially to the temperature of the earth. Nevertheless, even though Kelvin personally resisted the idea of transformation of species, his most limited calculations suggested an earth at least 5,000 times older than the Biblical age.

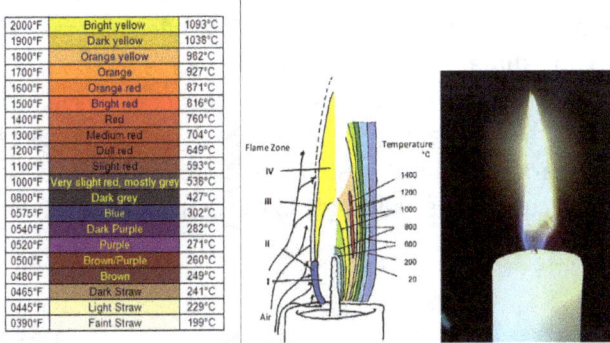

Figure 4.10. Left: Graph of color temperature[xiv]. Right: Color of candle flame[xv] indicates its temperature.

Evidence for the age of the earth: Red shift[xvi] and age of the universe

Everyone has noticed that, as a train passes by [xvii] or an airplane passes overhead[xviii], the sound of its motors changes (see a well-done Doppler effect animation and Doppler experiment or listen to the sound of an airplane[xix]). This is called the Doppler effect[xx], which arises because, if an object is approaching you at relatively high speed and is emitting sound, each sound wave initiates closer to you [xxi] and therefore arrives sooner, creating the effect of greater frequency or higher pitch. As it recedes from you, each sound wave initiates farther away, creating the effect of lower frequency or lower pitch. The same argument is valid for light waves[xxii]. Here, The difference is that, while an object has to be moving at a reasonable proportion of the speed of sound (600 miles/hour) for us to detect the effect, for light it has to be a reasonable proportion of the speed of light (671 million miles/hour) to be detectable and, of course, we have to know with substantial accuracy the base, or stationary, wavelength for the emitted light. Fortunately the light emitted by glowing elements provides us with this precision. When elements as opposed to compounds or impure materials are sufficiently heated or energized, they glow but only at very precisely-defined colors or wavelengths. You see this in the yellow color of high pressure sodium street lights, the bluish green color of mercury vapor lamps, the pure red of a neon lamp, and the pure and clean colors of light emitting diodes (LEDs). The purity of these colors can be used to measure time because of a phenomenon called Doppler Effect or red shift. If light is emitted from an object approaching you, the wavelengths will be compressed, moving the light toward the blue end of the spectrum[xxiii]. If the light is emitted from a receding object, the wavelength will appear to be longer or redder[xxiv]. It was a surprise to discover, therefore, that emissions of specific elements could be recognized from distant stars, but they were always displaced toward the red (longer wavelength) end of the spectrum. Such a result could come if all the stars were moving away from the earth, which could occur, for instance, if we were on the inside of a giant, expanding balloon. This was the primary argument that led to the hypothesis of an expanding universe. It was also possible to measure the speed of expansion by the magnitude of the red shift, and from that calculate when and where the expansion could have started. This calculation, a crude summary of the "big bang theory," leads to an estimate of an age for the universe of about 13-14 billion years, consistent with an age when our solar system could have formed about 6 billion years ago and the first appearance of our Earth 4½ billion years ago. Even though these numbers are subject to some revision, the likely revisions will be modest, and this calculation as well leads to the conclusion that the Earth is very old.

Evidence for the age of the earth: Radioactivity

The problem with most means of estimating the age of the earth is that there is no accurate clock. Accepting Steno's rules of sedimentation allows one to assess the order of events—that the age of amphibian preceded the age of reptiles, and the age of reptiles preceded the modern era of mammals, but one does not have an accurate measurement of rate. Sediment could be deposited at a gradual, consistent rate over several thousand years, or one massive flood might deposit an equivalent amount in a single event. It is possible to use the same rocks to argue for an age of the earth in billions of years or to posit that all the sediment was deposited in Noah's Flood a few thousand years ago. This is where the use of radioactive isotopes becomes so valuable. The most convincing independent means of assessing the age of the earth comes from our understanding of radioactivity works. Toward the end of the 19th C, Henri Becquerel and Pierre and Marie Curie discovered and began to work out the mechanisms of radioactivity. For our interest several properties of radioactivity are interesting:

Many elements come in more than one form. The chemical nature of the element is determined by the number of protons in an atom, or atomic number. Thus any atom that has six protons acts and looks like carbon and is so named, the element with an atomic number of 6. However, carbon can have a variable number of neutrons. Most carbon atoms have 6 neutrons, and hence an atomic weight of 12. More rarely, an atom can have 7 neutrons (carbon 13 or ^{13}C) or 8 neutrons (carbon 14 or ^{14}C); these three variant forms are called isotopes of carbon. ^{12}C and ^{13}C can be measured but are basically stable. ^{14}C, however, is a bit unstable and occasionally breaks down, with substantial interesting consequences. There is a trick to evaluating the breakdown of carbon though, and the measurement is easier to follow using lead (atomic number 82). Lead is mostly lead 207 (^{207}Pb, from the Latin name for lead) but also consists of several other isotopes. One of the isotopes, ^{206}Pb, comes from the decay of ^{238}U (uranium). Now the important issue is that the rate of decay is extremely constant. Chemical reactions, such as rusting, cooking, oxidation, mixing acids with bases, etc., depend on the interaction of two or more molecules, which move faster and therefore collide more frequently when they are warmer. Hence the rate of chemical reactions changes with temperature or concentration of the reactants. That is why cold preserves food and other things. Radioactive decay, however, depends only on the atom involved and is independent of temperature or concentration. If we have the radioactive isotope of hydrogen known as tritium (^{3}H) we know that half of it will decay in a little over 11 years, whether the tritium is found in ice, water, or steam or whether the hydrogen is bound to oxygen as in water or to any other atom, and no matter what its concentration is. The mathematics is a little more complex than a simple linear equation, since the constant is the time it takes half of what is present to decay (half- life). For instance, if we have 1000 atoms of tritium, in 11 years half will decay, leaving 500. In the next 11 years, half of the 500 will decay, leaving 250. In the next 11 years, half of those will decay, leaving 125. However, it is rather simple to rewrite our equations taking this into account. The important thing is that if we know the concentration (ratio of ^{3}H to the normal form, ^{1}H) we can calculate when it was 100% ^{3}H and when it will be effectively gone.

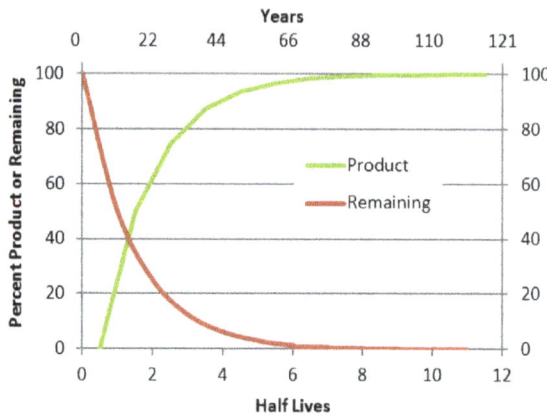

Figure 4.11. Decay of radioactive elements. The curve is logarithmic, with half of the remaining amount decaying in the time of one half-life, but can nevertheless be easily calculated.

This takes us back to our rocks. Suppose that we find a rock that contains a trace of ^{238}U, some ^{206}Pb, and other forms of lead. The mineral zircon is a good example. If we can measure the amount of ^{238}U and ^{206}Pb, we can calculate how long the uranium has been sitting in that rock, decaying to lead—in other words, the age of the rock. It so happens that the half-life of ^{238}U is very long, about 4.5 billion year (yes, it can be measured), so that if we find a rock that contains half ^{206}Pb and half ^{238}U, we can calculate that the rock is 4.5 billion years old. Rocks nearly that age have been found in Australia and Greenland, arguing that the earth is at least that old. Other isotopes have different half-lives, some much shorter than one second. Others however have half-lives more useful to us. ^{14}C, for instance, is formed only in the upper atmosphere from interactions with emanations from the sun. Once formed, typically as carbon dioxide, it drifts over many years through the atmosphere, finally reaching the surface, where it is likely incorporated into plants through photosynthesis. Once fixed in this manner, it will remain in the plant until the plant is fossilized or, if the plant is eaten, in the animal until it is fossilized. While some minor corrections are necessary because of recirculation and other minor inconveniences, by and large we can read the age of a fossil by the ratio of ^{14}C to ^{12}C. The half-life of ^{14}C is 5700 years. This means that we can read ages up to about 100,000 years before the amounts become too small to be reliable. We can use ^{14}C to trace a good part of the history of humankind, and we can use other isotopes to span other periods. Although radiodating was not well developed until mid-20th C, the point is that it is an accurate, reliable clock. Today we can state that there is no better explanation of the measurements obtained than that the Earth is a few billion years old. These measurements are completely consistent with totally independent measurements of astrophysicists, involving such phenomena as red shift. Measurements using shorter-lived isotopes also confirm quite ancient dates for dinosaurs and almost all other fossils.

Evidence for the age of the earth: Continental Drift

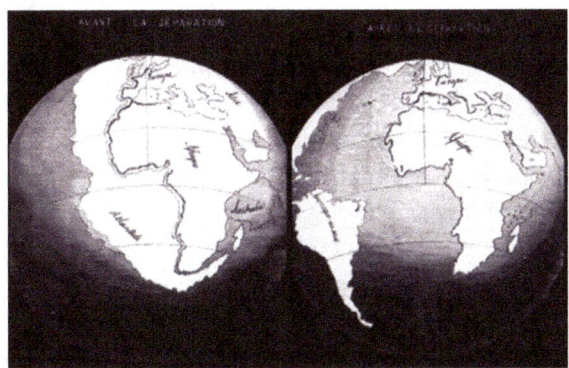

Figure 4.12. Snider-Pellegrini's diagram of potential continental drift. The left picture illustrates his hypothesized positions of the continents before the separation, and the right picture, after the separation.

Many a schoolchild has noticed the peculiar fit between the east coast of the Americas and the west coast of Europe and Africa. Cartographers had noted it too, as early as 1596, and in 1858, Snider- Pellegrini[xxv] had postulated that at one time the continents had been together and had later separated. However, it was not in anyone's realm of imagination that such an event had occurred at any measureable time in the history of the earth. During the 2nd World War, Harry Hammond Hess, a geologist and navy captain, used the newly developed sonar to map the Pacific Ocean during the maneuvers of his ship. He discovered and recognized undersea mountains and volcanoes. Using this information, together with the discovery by Bruce Beezen of the Great Global Rift[xxvi] (a recessed area between uplifts in the Atlantic Ocean), in 1962 he hypothesized that the continents were gradually drifting apart. Thus the mid-Atlantic Ridge, a volcanic undersea mountain chain that reaches the surface in Iceland, the Canary Islands, and Tristan da Cunha, is cut in the center by the Great Global Rift. He proposed that the hypothesis, called the Wegener Hypothesis (after Alfred Wegener, who in 1912 hypothesized that the separation was recent and continuing) was correct, and that the continents continued to separate along these ridges; in fact, that there were numerous continental plates, each capable of movement; that these movements were continuing; and that they had occurred in biological time. By the late 1960's, the U.S. National Science Foundation had launched a ship called the Glomar Challenger to explore the sea floor, and its first findings confirmed Hess' hypothesis. It had long been known that magnetic elements in the earth's crust can distort the magnetic fields, such that moderate corrections must be applied to compass readings, and the ship had set out to map them more completely. The researchers mapped the mid-Atlantic ridge. What they found was symmetrical deviations to the sides of the ridge: If there was a line 15 km west of the ridge in which the magnetic deviation from true north was 15° E of North, one could expect to find a similar deviation approximately 15 km east of the ridge. In fact, there was a series of stripes symmetrical about the ridge.

It is known that the magnetic poles have drifted relative to the configuration of the earth and its axis of spin. It is also well known that magnetized iron, when heated, will lose its magnetism but when cooled will assume the configuration of any magnetic field in which it finds itself, as the iron atoms resettle into position. Iron in lava routinely does this. The hypothesis therefore was completed to state that the undersea mountains were volcanoes, as are the islands, and that the lava flowing to the east and west of them took on the magnetic fields extant at the time of eruption. The only way in which would produce the patterns seen would be if the sea floor were in fact flowing from the ridge: each new eruption pushed the older material farther away. In this case, the [movement could actually push the continents apart](xxvii) (an excellent animation). It was possible, using radioisotopes, to measure the ages of the flows and to calculate how long this had been occurring. It turned out that the movements could be traced over a few hundred million years, with the Americas, Africa, and Europe fairly close together approximately 65 million years ago. Suddenly several biological curiosities now made sense. The existence of similar large, flightless birds on the three continents of the southern hemisphere had previously been thought to be an example of "convergent evolution," in which unrelated creatures evolve similar forms in response to similar conditions, like the marsupial (pouched) animals of [Australia that resemble rabbits, moles, and dogs](). Now it appeared (and was much later confirmed by DNA testing (in second volume[ixxviii]) that the birds were in fact related and had evolved from a common ancestor that roamed the southern hemisphere when the continents were connected. The existence of true apes (tailless, occasionally bipedal) and monkeys (typically with tails, and usually running on all fours) in the Old World but only monkeys in the New World could be explained by the argument that the monkey family had appeared and spread before the continents separated, but by the time the apes arose the gap was too great for them to cross. In terms of our question, the evidence that continents could and did move, and the evidence that this movement could be measured and gave figures in the millions of years, was further proof of an ancient world. Today, in a totally independent means of assessment, we can use satellites and lasers to track what we today term continental drift. Its rate is between a few inches and a few feet per year, consistent with earlier measurements. The hypothesis of continental drift also explains mountain building and earthquakes: As the plates that carry the continents collide, one typically slips underneath the other, lifting its edge into mountain ranges such as the Andes, the U S Coastal Range, and the Himalayas. Where the first plate goes down ("subducts") it dives deep into the crust, opening channels through which lava can rise. The movement of these plates generates the "[ring of fire](),"[xxix] the volcanoes that circle the Pacific Ocean.

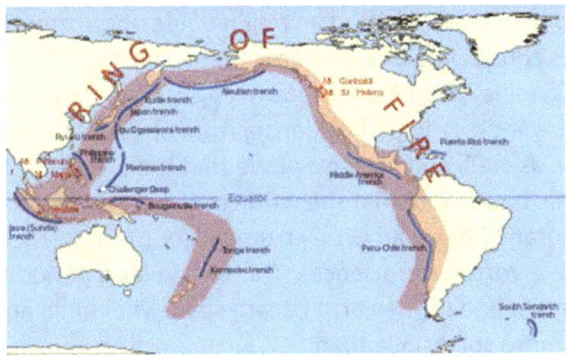

Figure 4.13. The ring of fire.

Recognition of the reality and timing of continental drift and of the existence of the "dinosaur comet" represent two more marvelous detective stories, again illustrating why to be a scientist feels like forever learning the tricks of a magic show. We will tell these stories in volume 2^{xxx}.

DNA

Today we can even measure time by DNA. This sounds weird, but it is not so complicated. Most people today have heard that DNA consists of a string of four types of "bases" and that the order of these bases is so elaborate that we can determine identity of a person from this order in the same sense that a sentence or paragraph (a specific sequence of letters) can be unique. Imagine the DNA as a long sequence of "snap together" beads, each of a different color, such as are used for children's necklaces. Imagine further that you swing it over your head until the string pulls apart and splits into two. Imagine also that when you put it back together you can snap the "plug" end back into the hole from which it came, or you can snap the plug into the other free "hole" end. If you keep doing this, sooner or later you will have scrambled the order of the beads (Fig. 18). If you posit further that, on average, the string will break at a specific frequency, you can determine both the time since the first break and the order of breaks. This is what is done to determine when specific species became distinct from each other (for instance, when the protohuman line separated from that producing the great apes) and the relationships among different groups (for instance, the type of mammal to which whales are most closely related). We'll deal with that later in considering how and when these groups evolved, but in this context we establish a great age for the existence of life on this planet, over two billion years, in fact. And the important point is the convergence of many different types of data— multiple independent sources—to a single approximation of the age of the earth. EVERYTHING—astronomical measurements of red shift and other activities; radioactive decay; calculations of the rate of cooling of the earth; measurements of continental drift and other geological processes; evidence for erosion, sedimentation; calculations of the rate of change of DNA; and many more measurements—points to the great age of the earth, giving it an origin approximately 4 billion years ago and an origin of life approximately 2 ½ billion years ago. It is this confluence of arguments that is most convincing. If we were to insist that any single line of evidence were wrong, we would be obliged to find fault with every single other line of evidence, from physics, astronomy, geology, molecular biology, as well, and likewise prove that the principles that guide these sciences are wrong. If we insist that the writings of our holy works (Hebrew Bible? Christian Bible? Koran? Bhagavad Gita?) trump all of these sciences, we still are obliged to find the errors in the sciences. The 19th C was a period in which evidence for the great age of the Earth became overwhelming, and the evidence has become only more solid since then.

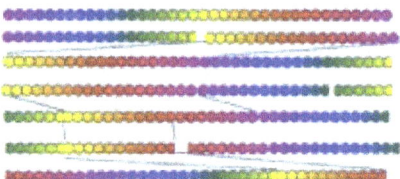

Figure 4.14. Measuring time by DNA. Assuming that DNA strands break on average once every X number of years, by calculating the number of breaks from an original strand (top) one can estimate the number of years since the origin of the sequence. The sources of the original pieces are established by sequencing DNAs of related species and constructing a hypothetical original DNA from the recognized fragments (indicated by linking lines).

Conclusion: The evidence for an ancient Earth is overwhelming

There are of course errors associated with all of these measurements. In addition to problems caused by human addition of radioisotopes to the environment, when we measure vanishingly small amounts of materials we risk contaminating the specimens, and other sources of contamination such as water seepage, growth of roots, and movement of dusts must be accounted for. Thus many estimates of age that were once published with confidence have subsequently been corrected and revised. Nevertheless, the results are clear and confirmed by measurement of several different isotopes: The earth is approximately 4.5 billion years old and for perhaps 3 billion years has been cool enough to support life. There has been plenty of time for evolution to take place.

The point is that it would be fair enough to pit Darwin (actually Lyell) against the writer(s) of Genesis concerning the age of the Earth, stating that it was the word of one person against another, in which case we could reduce the tension to deciding that Divine writings pre-empted logic and evidence or vice versa. However, should we deny Lyell's argument, we would then be obliged to find flaw with the logic and evidence collected by scientists studying geology, astrophysics, atomic energy, chemistry, geophysics and many other sciences. For each, we would have to find reasons why their logic was faulty, and erect new hypotheses to explain the phenomena on which the hypotheses were based. The entire structure of science would founder. When a hypothesis is built on several completely independent but consistent bases of measurement, denying the hypothesis requires dismantling an entire superstructure of scientific reason, taking with it the explanations of how many aspects of our modern life work, from calculating the trajectories of missiles to generating power in green or non-green methods to designing and using incredibly sensitive sensors, medicines, and electronic devices.

The question of understanding time was one of the fundamental blocks to recognizing that evolution had occurred and could be explained by recognized biological processes. There were two other intellectual blocks. One was that thinkers had to accept the idea that *species differed across the face of the earth*, and for no obvious reason. The second was to acknowledge the *cruel logic of nature*[xxxi].

~~~~~

Chapter 5: A question of mechanism—the driving forces of evolution

Science and hypotheses are about mechanisms, whereby one cause produces a specific effect. By 1850 at least some people agreed that the earth was old, that there were startling resemblances in the structures of animals, that fossils were the remnants of real animals that were different from modern-day animals and that, the closer in the fossil record one got to the present, the more closely the fossils resembled modern creatures. In general, animals and plants seemed well adapted for their life styles. People also agreed that, contrary to what one might assume, animals and plants from different continents differed in unpredictable ways. But the arguments were intense as to why the world should be ordered in this fashion. It was also clear that humans could change domesticated species. What was missing was a mechanism. This was where Darwin and Wallace entered the picture, and here they needed the help of the social scientists, especially Malthus.

Urbanization, Industrialization, and the lessons of Malthus

Science is about how things work: what mechanism or force causes or determines a specific change. The mechanism was to come from a cold-eyed and pessimistic man of God who wrote a book. Both Darwin and Wallace read the book, <u>Essay on the Principle of Population</u>.[i] Darwin read it two years after returning from the *Beagle*. Both biologists had accepted the ideas of the early 19th C, and both recognized in Malthus' arguments a mechanism that could drive evolution. In this manner the conversion of several ideas led to the apparent coincidence of two scientists almost simultaneously coming to the same conclusions. There are many such other coincidences, such as three laboratories <u>rediscovering Mendel after 32 years of neglect</u>, the simultaneous development of three versions of a polio vaccine, the race to identify the genetic material, the identification of messenger RNA[ii], and many others; they similarly result from the coalescence of several trains of research.

What had happened was that social inequity had been recognized as a problem, and people were beginning to think about it. Life was not very good in big cities as the Industrial Revolution got underway. In hunter-gatherer societies, provided that conditions are stable, life is not bad. People have to work far less than urban sophisticates to feed themselves—3 to 16 hours/week—their food is varied and nutritious, and most forms of pollution are relatively low. As cities grew, they were crowded, filled with rats and vermin, contaminated with human and animal waste, with polluted air and water, and conducive to the rapid spread of disease. The deteriorating health of urban populations in the late 18th and early 19th C can be easily documented by church records listing births and deaths including infant deaths, size and condition of skeletons recovered from graves, and written and physical evidence (skeletal or age at parenthood) of age at puberty.

Most critical was the availability of good and varied food. This latter issue is easy to imagine if you consider a city like New York City. It has an urban population of approximately 8 million and a conglomerate population of approximately 20 million. Essentially no food is grown within the conglomerate region. This means that enough food for 20 million people has to be brought into the region every single day, and the waste removed. If you put this in terms of the caloric value of steak, you would require 14,000,000 pounds of steak every day, or 27,000,000 loaves of bread. In terms of milk, it would be about 18,000,000 gallons of whole milk or 7-15 tanker trucks every day. This is a pretty big job. Now imagine that you have no refrigeration, no motorized vehicles, and little ability to prevent depredation by rats or other vermin. The problem becomes more obvious. In the early 19th C, food had to come from areas within the range of a burden-bearing animal and it had to remain edible for a reasonable length of time. In winter, one would be limited to those grains and tubers that could be stored, and to meat and eggs. As cities grew, it would be harder and harder to deliver enough food to them, particularly insofar as physical expansion pushed the growing areas further from the core of the city.

And therein lay the problem. The economist Reverend Thomas Malthus (1766-1834) put together and quantified what was intuitively obvious: people have more children than are needed to replace them. First, suppose that a couple begins to have children at age 20 and husband and wife live to age 80. Even if they have only two children and each generation has only two children, just before the parents die there will be ten people, counting children, grandchildren, great grandchildren, and great great grandchildren. However, if each child (not each generation) has two children, by the 80th year there will be 32 people. If each couple bears 3 children, by the 80th year there will be 81. And humans frequently have more than three. Where expansion is unlimited, as was the case when English settlers arrived on the Falklands Islands (Islas Malvinas), they might average eight children/couple. This is a logarithmic projection, in which the number increases as a multiple of the number of iterations. If the number doubles each time (n*2, each individual produces two children, or each couple produces four children), then the number per generation would be 2, 4, 8, 16, 32, 64, 128, 256, 512, 1024...a thousand-fold increase in 10 generations. Malthus pointed this out forcefully, counting births, marriages, and deaths between 1692 and 1716. He found that births always exceeded deaths, usually by a factor of 1.6 births for every death. In 1709 and 1710 plague increased the number of deaths from around 8,000/year to 125,000/year. After the plague subsided, the number of deaths dropped to around 10,000/year but the number of births increased to over 32,000/year, for a ratio of 3.2 births per death, and it remained almost 2:1 for the next five years. Thus he documented a logarithmic increase in population. However, he feared that food production would increase only linearly, 1, 2, 3, 4, 5, 6, 7, 8... He therefore argued that there had to be a struggle for existence in which people would compete for food and not everyone would survive. He predicted a catastrophe by the middle of the 19th C. His essay is available online at http://www.econlib.org/library/Malthus/malPlong.html (6th Ed)[iii] and http://www.econlib.org/library/Malthus/malPop.html (1st Ed)[iv]; it is short and eminently readable.

To be fair, he underestimated the ability of people to invent new ways of improving crop yields, preserving food for storage, and transporting it. Although there certainly was wretchedness, as described by Dickens, Zola, and others, and social turmoil followed, it was not quite as bad as he predicted. Nonetheless, his calculations and warnings remain valid. Although Europe has increased its food production per capita over the last 50 years (mostly because its birth rate has fallen), Africa has not,[v] and that of Somalia has taken a substantial drop.
[vi]Malthus' warnings were dire: Since populations will outgrow their food sources, there will be struggles and some will die. Only the strongest will survive, limiting the growth of the population. Both Darwin and Wallace read the book. Wallace was in the Far East, and he called it "the most important work I read". Darwin read it as a diversion while he sorted through and tried to make sense of his collections; he called Malthus "that great philosopher". Thus was born the first coincidence in the history of evolution. Both Wallace and Darwin understood immediately that Malthus' calculations applied directly to the biological world and that this competition could provide a mechanism for change.

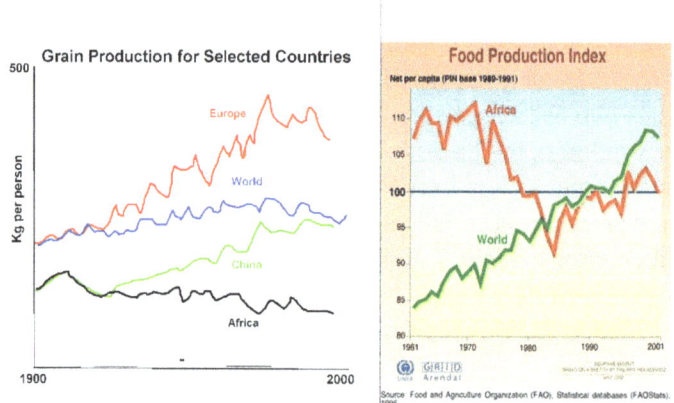

Figure 5.1. World food production in kilograms per person[vii]: Left: From top to bottom: Red: Europe; Blue: World; Green: China; Black: Africa. Right, a different chart indicating food production in Africa vs the world[viii]. Although at present total world production is keeping up with population growth, Africa has fallen since 1960, and the expansion cannot continue indefinitely.

In retrospect, the Malthusian calculation (logarithmic increase in population, linear increase in food supply, consequent crash in population) was obvious if anyone had bothered to think about it. Many fish can spawn 10,000 or more eggs, and trees can shed over 100,000 seeds. Even out of the spectacular realm, most insects produce a few hundred eggs, birds often lay 6 or more eggs/year, and many mammals have litters of 5-10 pups. Horses were introduced into Argentina in the 16th C. There were 12 to 45 wild (escaped) horses in 1540, and 12,000 by

1580[ix]; there was a similar expansion of goats on the island of St. Helena[x]. Yet none of these species has overwhelmed the earth. Their numbers appear to be more or less steady. If there are 20 fish (10 pair) in a pond, and they lay a total of 100,000 eggs, and the following year there are still 20 fish, then at least 99,980 of those eggs must have perished. Over 100 years an oak tree in a forest may produce 10,000,000 acorns. If at the end of those 100 years the forest has not changed much, by definition 9,999,999 of those acorns never made it.

Now, at last, we have a mechanism. The mechanism could have been identified much earlier, since even pre-literate humans were aware that many animals and plants produced far more young than would have been necessary to maintain the population and even in the earliest agriculture it is obvious that the practitioners would have determined and limited how much seed they needed to retain for future crops as opposed to how much they could afford to eat. But the recognition of its importance arose only in the 19th C, owing to the increasing poverty in the cities. Malthus published the first edition of An Essay on the Principle of Population in 1798, describing the "struggle for existence"; Dickens' novels were written primarily in the 1850's; and Herbert Spencer began to toy with the ideas that were later called social Darwinism as early as 1851. Spencer read the first edition of Darwin's *On the Origin of Species* and first described the process as "survival of the fittest," a term that Darwin used in later editions.

The Mechanisms

The mechanism that both Darwin and Wallace both saw derived from Malthus' observations on human society:

- All species produce far more young than would be need simply to replace themselves.
- Nevertheless, most populations tend to remain constant in number.
- There are never enough resources (food, suitable nesting sites, or other resources necessary for survival) to support a substantially larger population.
- Therefore, there must be a competition or STRUGGLE FOR SURVIVAL among individuals of the same species, of which only enough survive to replenish the population.
- In any species, there is variation among individuals in all aspects of their appearance and behavior. (Here Darwin embraced variation not as a flaw in perfect descent but as a natural and necessary aspect of his argument.)
- If for any reason some variations make it more likely that the individuals survive, these individuals will be the ones to produce the next generation. (Note that any variation might do. It could be, for instance, a flatter cockroach that can squeeze more effectively into a crevice, or a

color closer to that of the background, or the ability to withstand a mild frost. Darwin is going to call these variants more fit, but by this term he refers only to the ability of the variant to leave progeny to the next generation, not to any physical strength.)

- If only the most fit individuals produce the next generation, gradually the species will change so that the species as a whole will more closely resemble the most fit individuals. This is DESCENT WITH MODIFICATION.
- Eventually this modification will become so extreme that we will recognize a new species. This is the ORIGIN OF THE SPECIES.

Everything has come together: The evidence is that there is variation within a species, that species vary around the world, with different but similar species living in geographically related areas, and that species have changed in the history of the world; that the world is old enough to have supported not only minor but profound changes in species; and that there is manifestly a struggle for existence, in which only a few survive. The mechanism hypothesized is that this competition selects variants, such that the species as a whole becomes more "fit," or better capable of meeting its major challenge, so that the species changes with time, gradually creating new species. This is the primary hypothesis: NATURAL SELECTION. Darwin compared it to the selection consciously performed by animal and plant breeders, whereas Wallace thought that the two processes differed.

Before the question arises: species do not become "perfectly fit" for several reasons, among which three are very important. First, conditions change reasonably rapidly, so that, for instance, a color that is good camouflage in one setting is not as good in a subsequent setting. Second, the evolution of one species does not occur in a vacuum. If a rabbit evolves to outrace a dog, the dog must evolve to become faster or more capable of surprising the rabbit, or it will starve. Third, being "perfectly fit" creates a problem in that if the population overbreeds, it will eventually strip its habitat of all resources. Quite frequently, if a species invades an island in which it has no predators, it will fill the island, and the population will subsequently crash because of disease or starvation. Sometimes it is better for a species to tolerate a certain level of predation to keep the population under control than it is to reproduce without limits. The number of hare builds in the arctic, followed by an increase in the population of lynx; the increase in predators cuts back on the hare population, which leads to a decline by starvation of the lynx, which allows the hare population to build… An effort to expand the deer population near the Grand Canyon by killing wolves and coyotes was very successful—for a few years. Soon the deer population had quadrupled, whereupon the deer ate all available vegetation by midwinter, and starved to death. The population crashed to 1/10 of its original size, and took many years to rebuild to where it had been.

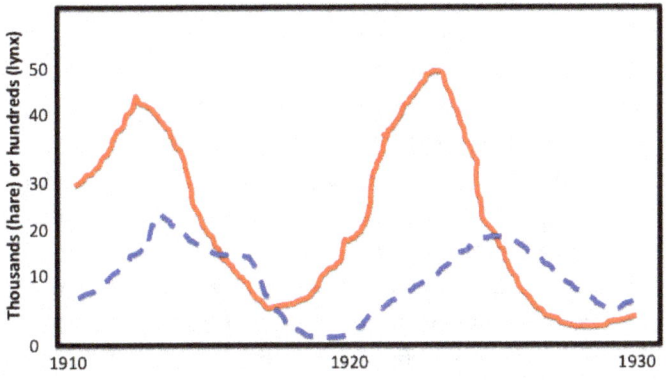

Figure 5.2. Population of hare (red, in thousands) and lynx (blue, in hundreds, in Canada, as reported by MacLulish in 1937.

A question of variability--The biological world: Were there kangaroos in Noah's Ark?

One of the first questions that began to trouble explorers were the different creatures that they encountered, because the existence of these creatures was not easily reconciled with their understanding of the relationship of animals and plants to humans. Columbus knew that he had reached islands, not a continent, because there were no large mammals present—but why were there no large mammals on planets? Dogs, for instance, were meant to serve man by defending houses and territories. However, the natives of the Caribbean kept as pets dogs that could not bark. How could a dog that did not bark protect its master? How did such a dog fit into the role of dogs *vis-á-vis* humans? More complex, God had created earth for the pleasure and utility it afforded mankind[xi]. However, those humans blessed by the knowledge of God inhabited Europe, and there was much to desire in the New World: fur-bearing animals such as raccoons, opossums, and skunks; turkeys; food plants such as corn, tomatoes, potatoes, sugar cane (known but hitherto ignored), sunflowers, and chocolate; true cacti; and many types of flowers (we know today that glaciers had pushed flowering plants out of Europe, and to return they would have had to cross the Sahara and the Mediterranean, whereas in North America they simply re-expanded from the South). Even poison ivy was originally considered to be an attractive houseplant. Europeans began to question why God had given savages such beneficence, when Europeans would have enjoyed these creatures. More importantly, the bizarre distribution of creatures could not be explained by the story of Creation and the Flood. Why were creatures found on one continent and not another? One could understand why there were no lions in Europe. Lions could walk into Europe from the Middle East, but they chose not to; it was too cold or otherwise not desirable to set up camp in Europe. However, the climate in North America was similar to that in Europe, but many European creatures were nowhere to be found. Presuming that all these species had been on Noah's Ark, how could they have ended up on different continents? Surely Noah's Ark was not some sort of school bus, dropping off animals and plants on different continents. (The heading is a bit unfair, since kangaroos were not really seen by westerners until the early 19th C, but the point is valid for less spectacular creatures.) And furthermore, when known animals such as gulls or squirrels were found in new continents, why did they differ in appearance, song, or habit from those known in Europe? Was each variety represented on Noah's Ark, meaning that they were different kinds of animals or plants? Or did they all represent only one kind with variation? What in fact did we mean by "kind"? Such questions expanded, culminating with the attempt of Linnaeus in the 18th C to name and classify all known species. This question became a driving force in the effort to understand the structure of the biological world, and ultimately to question the stability of a species, as is described in Chapter 6.

The biological world: What is the relationship of one organism to another?

By 1735 the number of known animals and plants had grown so large that Karl Linnaeus attempted to group them into rational categories. Such efforts had been undertaken before, and it is even reasonable to assume that all societies, and even everyone, makes some sort of unconscious grouping. After all, we all recognize mammals, birds, frogs, and fish; we recognize that insects, lobsters, spiders, snails, and worms are different; and we readily subdivide mammals (carnivores and herbivores; cats, dogs, rodents, etc.) We sometimes make mistakes: Is a bat a bird or a mammal? Or a tree might be more closely related to a shrub than to another tree. But, by and large, we usually have no trouble. In fact, until the 18th C, Europe mostly relied on the seemingly sensible classification of Aristotle (with or without blood = vertebrates vs invertebrates). In fact, Aristotle classified *everything*, including the earth. His hierarchy, according to one compilation, was as follows (the list was drawn with an indicator going from bottom to top and marked "Tending toward perfection":

- God
- Humans
- Mammalian animals
- Flying squirrels
- Bats (and birds?)
- Fish
- Reptiles
- Shelled animals
- Insects
- Sensitive plants (for collapsing fern see LINK[xii]
- Plants
- Short mosses
- Mushrooms

- Stones
- Crystalline salts
- Metals
- Earth

Today we recognize the obvious mistakes. Humans and flying squirrels are mammals, as are bats. Sensitive plants are plants, even though they react (by collapsing leaves) when disturbed. Linnaeus was trying to correct this table and make it more comprehensible by subdividing groups. For instance, among birds there are the predatory birds such as hawks, falcons, and eagles; shore birds such as gulls; ground birds such as chickens, grouse, and quail; duck-like birds; penguins, etc. He also subdivided the subdivisions. Insects include beetles, dragonflies, butterflies, grasshoppers, bees, etc. The group butterflies includes butterflies and moths; butterflies includes those with tailed wings (swallowtails), cabbage whites, and the iridescent tropical Morphos, and so forth. He described these relationships in a subordinate structure, as we might describe plants>broad-leafed plants>trees>fruittrees>citrus trees>orange tree or, disregarding the artificiality of political boundaries, humans>Europeans>Italians>Napolitanos>Maria. To make these subdivisions he created a hierarchy of classifications, and therein lay the problem. To Aristotle there was an order or rank to nature, in which everything had a specific location. Thus a cat had to be above or below a dog, not equal to it. A sheep had to be above or below a goat. Among the cats, lions, tigers, leopards, and housecats all had an appropriate, and fixed, position. To Linnaeus, the situation was more complex. (There was also another problem. To Aristotle, variation was deviation from perfect. For every species, there was an ideal form or ιδεα (in fact, this is the origin of our words 'idea' and 'ideal'). All creatures aspired to that idea, coming more or less close to it. More about that later.

The first problem that Linnaeus encountered was the existence of parallel paths. Why should birds be higher or lower than mammals? Why not at an equal level? Among the scaly-skinned animals, there were turtles, snakes, lizards, and crocodilians. They were all clearly similar in many respects to each other, and should be classified separately from other animals, but not necessarily above the slimy- skinned animals (frogs, toads, salamanders, and newts). He then subdivided the categories into finer and finer detail, much like defining a location in an encyclopedia: *Encyclopedia Brittanica*, Volume "M", topic "mouse", paragraph "field mice," subdivision "deer mice"...

He made a few errors, and he had a series of unclassifiable creatures, listed as paradoxes, that he did not know how to handle, including pelicans and many mythological creatures, such as satyrs, borometzes (plants that were supposed to grow sheep as fruit), phoenixes, and dragons, which he had no reason to assume did not exist, but basically his groupings and subgroupings were as follows (put into modern form):

Domain: Eukaryotes (all living organisms whose cells have true nuclei, as opposed to bacteria and viruses)

Kingdom: Animals (all organisms that feed and do not rely on photosynthesis or chemical degradation of material, such as all plants and fungi)

Phylum: Chordates (all animals with a backbone or, strictly speaking, with a structure within the backbone, the notochord. This includes all true vertebrates and sharks, but leaves out molluscs, worms, insects, crustacean, etc.)

Class: Mammals (all warm-blooded chordates with fur and which bear their young live and suckle them with milk; as opposed to birds, fish, reptiles, and amphibians)

Order: Carnivores (mammals that eat flesh and have sharp teeth that can tear flesh; not rodents, bats, whales, or grazing animals)

Family: Cats (cats, lions, tigers, panthers; not dogs, skunks, or otters)

Genus: Felis (small cats including domestic cats and similar wild cats; not lions, tigers, or lynx)

Species: cat (the domestic cat, differing from jungle cats, mountain cats, sand cats, and wildcats)

(Two generally accepted conventions are that we usually refer to animals or plants by their genus name (capitalized) and species name, as in *Felis catus*, and for general purposes we define a species as a group of organisms that can interbreed. Thus, all varieties of humans can produce, in any combination, normal, healthy, strong, and fertile children and are therefore one species. Horses and donkeys can mate and produce a healthy, strong offspring (a mule) but it is sterile and cannot contribute to the furtherance of horses, donkeys, or mules. Thus horses and donkeys are different species. In spite of their enormous variation, dogs are considered to be the same species because, although it is wildly impractical, it is theoretically possible for the genes of a Chihuahua to mix with the genes of a Great Dane (through several generations of intermediaries of appropriate sizes). The large variability is characteristic of domestication.

Figure 5.3: Earthworm, velvet worm,[xiii] centipede, northern[xiv] and southern leopard frogs[xv]

Figure 5.4: Variety of butterflies of the same species[xvi].

The second problem that Linnaeus encountered was that his very effort undercut the philosophy of what he was trying to do. He considered that he was doing God's work in identifying a placing in order every type of creature that existed on earth and had climbed on to the Ark. However, the greater effort that he invested in this project, the less clear the separations became. There were soft-skinned segmented animals (earthworms or Annelids—the Latin word means "ringed") and hard-skinned segmented animals with legs (millipedes and centipedes—thousand-legged and hundred-legged, belonging to the group Arthropoda, or jointed-legged animals). Fair enough. But what about a creature described from Australia, a soft-bodied segmented animal, with soft, fleshy legs and antennae? Was it an annelid or an arthropod? And sometimes individuals of one species varied among each other, with some individuals so much resembling another species that it was very difficult if not impossible to tell where one species ended and a second species began. Linnaeus' Herculean effort was so well-done that it ultimately led to the failure of Linnaeus' goal. He had presumed at the outset that he was categorizing and systematizing all creatures that existed on earth and had therefore been on Noah's Ark. It was a more gigantic task than biblical or medieval scholars had expected, but it was finite and doable. However, the greater detail that Linnaeus and his followers achieved, the less certain it became that closely-related species could be separated. If two species of butterflies were orange and brown, respectively, there might be very dark orange variants of the first species and very orange-ish variants of the second. It would take an expert to identify the species of each, and another expert might dispute that claim. This became the basis for amateur and professional collections of variants of species, commonly found in museums today (Fig. 21). To take another, well-known, example from today: _Leopard frogs from Quebec_ are noticeably different in appearance from leopard frogs of Louisiana. Their calls are very different[xvii], and they do not spontaneously interbreed. If one takes sperm from a frog from one location and uses it to fertilize the eggs of a frog from the other location (in frogs, males squirt sperm onto eggs as they are laid, and so the process is easy to replicate in the lab), the eggs do not develop into viable tadpoles. Thus by this criterion the northern and southern frogs are different species. However, if one crosses a frog from Quebec with a frog from New York State, they interbreed, as does the New York frog with one from Pennsylvania, the Pennsylvania frog with one from Virginia, the Virginia frog with a Tennessee frog, the Tennessee frog with a Georgian frog, and the frog from Georgia with a Louisiana frog. Thus by this criterion they are the same species.

The point is that, the closer one looked, the harder it became to define the boundaries between some species. This problem is extremely perturbing to someone who considers the species to be fixed and distinct, as was assumed by Aristotle's _idea_ or the authors who chose to put into writing the story of Noah's Ark. It almost looks as if the concept of species is malleable, and that species are in fact not fixed.

Anatomy and fossils

The problem was compounded by the increasing interest in anatomy and in fossils. In particular, the story begins with Georges Cuvier, the director of the Musée d'Histoire Naturelle in Paris on and off until his death in 1832 and holder of several prestigious professorships and titles during the first third of the 19th C. He started with the basic and obvious observation that one could understand the lifestyle of an animal from its anatomy. Whether you realize it or not, we all can do this. Predators have strong claws, sharp, tearing teeth, and forward-facing eyes so that they can judge distance accurately. Herbivores have eyes to the sides, so that they can see well from front to back, and they have flat, crushing teeth. Unless for special circumstances they need to dig, they do not have slashing claws. These rules apply from praying mantids to lions, and from grasshoppers to cows. A movie that portrayed a guinea pig-like animal as a dangerous predator would be ridiculed. Dangerous creatures in science fiction have the requisite claws, teeth, and forward-facing eyes. Going further, strong flying birds typically have keels on their sternums, a projection of bone in the plane that divides the left and right sides of the body, because that is where the flight muscles attach. Skeletal muscles pull in a linear direction, so to pull the wing down and thus lift the bird up, the muscle has to attach between this keel and the upper shoulder. We cannot fly because we have flat sternums to which our pectoral muscles attach and we cannot get the force to lift our weights. There are many such observations to be made. It is obvious that sea lions, with front flippers and rudimentary hind legs, almost converted into flippers, will be far more agile in water than on land.

Figure 5.5 Related herbivores or scavengers (top) and carnivores (bottom). From left to right: Top: <u>Plague locust</u>, bullfrog tadpole, mourning dove, capybara. Bottom: <u>Praying mantis</u>[xviii], green frog, golden eagle, cheetah. Herbivores have eyes set to the sides, so that they can see what is approaching them, while carnivores have front-set eyes, for depth perception. Small carnivores like a mantis will sway from side to side to get better estimates of distance.

Figure 5.6. Adaptations for flight. Left: Birds like this chicken have large keels (sternum, black arrow) to which the powerful flight muscles are attached. Ground birds frequently have very large ones because lifting off the ground requires great force. Middle: A <u>bat</u> is a mammal but even so, it has an expanded sternum (black arrow). Most mammals (right: cat, from front) have a flat sternum (white arrow).

Cuvier followed this logic quite thoroughly, extending it to the relationship of each part of an animal to another and realizing that he could extend it to fragmentary information about an animal. For instance, the pelvis of a true quadruped (cow, horse, dog) links the horizontal spine to the vertical hind legs and therefore the joint where the spine attaches must be at right angles to the joint for the legs, whereas for a human pelvis the joints must more-or-less align (Fig. 5.6) Even the femur (thighbone) forms a ball that fits in the socket of the pelvis, and one can read from the angle of that ball the position of the pelvis. Seeing teeth would allow one to predict the type of feet (predators would have sharp teeth and slashing claws; herbivores would have blunt teeth and hooves). Similarly, our thoraces are barrel-shaped (pix) whereas those of apes (and most of our ancestors) are more trapezoidal. The narrowness at the upper end allows room for much larger shoulder and breast muscles, needed for them to be able to spend most of their lives climbing trees. All efficient swimmers have similar shapes, as do efficient flyers, since propelling oneself through water or through the air imposes more physical limitations than does lifting one's weight from the ground (you cannot hop, like a kangaroo, grasshopper, or springtail, in water or air).

Thus Cuvier was reputed to be able to reconstruct an entire animal from a single bone. He turned these talents to the examination of fossils that were attracting more and more attention. He examined skeletons of mastodons and giant ground-dwelling sloths from the Americas, as well as the fossils of molluscs. He recognized that the fossils represented real animals that had once walked the earth and had had real lives but no longer existed. (Others had assumed that fossils represented deformed known species or artifacts placed by God or the Devil for any of several purposes.) Simply put, Cuvier insisted that some animals has once lived on earth but had gone extinct. Furthermore, he recognized that the stratification of fossils indicated a sequence of change. In 1812, he published an important book entitled *Research on the fossil bones of quadrupeds, from which one reestablishes the characteristics of several species of animals that the upheavals of the earth appear to have destroyed.*

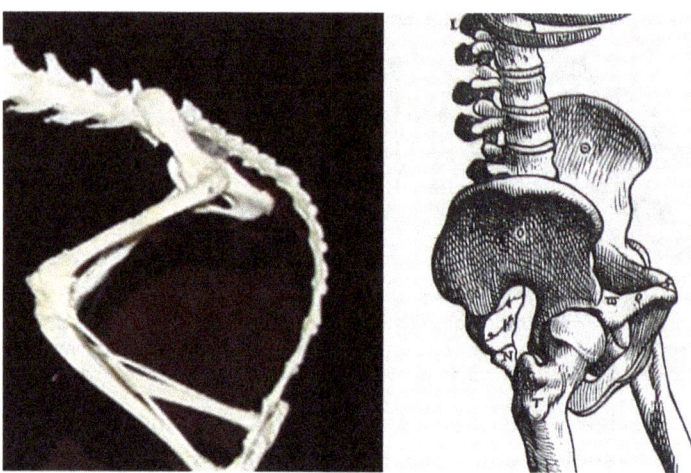

Figure 5.7. The shape of the pelvis indicates means of locomotion. Left, in the pelvis of a cat, the spine and the hind legs form an acute angle. If one were to stretch out the cat (have it stand on its hind legs) it would be obvious that the fit in the pelvis is very awkward. Right: In contrast, in humans the spine is almost parallel to and directly above the legs, and the pelvis is rotated and expanded to support the viscera (which are supported by skin and muscle hanging from the spine in a cat).

And yet, he is not the father of evolution. He had no concept of the age of the fossils and estimated the time involved as representing a few thousand years, as opposed to the hundreds of thousands of years that Lamarck and others were suggesting—that they invented their timetables "with the stroke of a pen". Furthermore, his belief in the integrity of the anatomy prevented him from accepting the possibility that modest, coordinated changes could occur; he considered that any single change would have to be detrimental, and that each variety had to be a new creation.

He was renowned for identifying and interpreting many fossils, including those of mastodons and giant ground sloths. He recognized that these New World animals were very similar to extant species (elephant and sloth, respectively but, in addition to being much larger, they differed in many respects from modern animals. Most of the skeletons were not found intact, but he had a very careful eye, and was reputed to be able to reconstruct an entire animal from a single bone. (This is not as astonishing as it might seem. For instance, an animal with slashing teeth, like a lion, is most likely to have powerful running legs. The hind legs of a quadruped articulate very differently to the pelvis than the hind legs of a biped like a human. A bird or a mammal with fin-like legs, like a penguin, otter, or orca—see Fig. 9.7— is likely to have other adaptations for life in water.) In any event he did not deny the claim. The problem was twofold: First, he could clearly identify animals that functioned and lived in times past, but for which he had no modern understanding. A pterodactyl, for instance (Fig. 5.8) was clearly a soaring creature, but with reptilian features, unlike any known reptile. Second, what we would today call homologies, similarity in bone structure among all the vertebrates, provided him with a sense of order and a means for identifying and organizing his fossils but created a very perplexing situation. As his mentor Geoffroy Saint-Hilaire had famously declared, "There is only one animal," arguing that the bones of all the vertebrates were essentially the same. But why should this be? In today's terms, we can easily recognize many of our body parts that an engineer, starting from scratch, might have designed differently. For instance, the photoreceptor cells, the cells that receive light in our eyes, are turned inward, with their light-sensitive parts facing inward, and their tails, connecting to the nervous system, facing outward. This limits the resolution of our eyes. The broad S-shape of our spine is not the most efficient means of balancing our entire upper body on our pelvis, and is the source of many backaches. To take the spine as an example, why adapt for two-legged purpose the arched spine suitable for suspending the body of a quadruped between four legs (Fig. 29), rather than designing it anew? Debates over what God meant or intended by arranging such similarities were an important feature of the intellectual ferment of the early 19th C. Cuvier, in fact, had considerable trouble with this. Although he saw differences in animals in moving from one stratum to another, with growing resemblance to modern forms the higher he got, since he never identified a creature that he could truly consider to be an intermediate between two forms, he did not believe that there was any familial link between the forms, or what we would today call evolution. (The absence of intermediate forms is the basis for a modern interpretation of evolution, called punctuated equilibrium, promulgated by Stephen Jay Gould, which argues that most evolution occurs rather rapidly in brief periods of very high selection pressure, followed by longer periods in which the species is relatively stable.) In our context, however, the point is that this ferment existed. Fossils were known, recognized, and acknowledged. What they meant in terms of the structure and history of the world was an open and important question

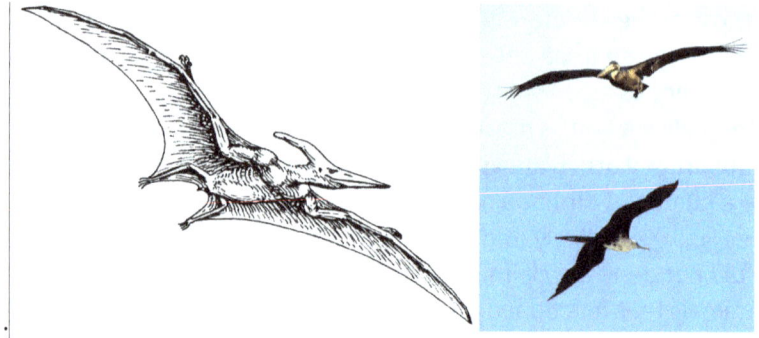

Figure 5.8. Left: Image of a flying pterodactyl[xix]; Right: a pelican (above) and magnificent frigate bird (below) in flight. The pterodactyl closely resembles the shape of these soaring marine and coastal birds, and must have had a similar lifestyle. (Birds that soar over land, such as vultures and eagles, typically have broader wings.)

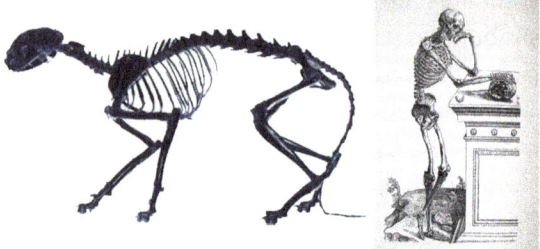

Figure 5.9. The spine of a cat is arched, a strong weight-bearing structure, whereas that of a human has acquired secondary curvatures.

The findings of the anatomists were intriguing and raised another, more complex question. The skeletons of all the vertebrates resembled each other to a remarkable degree. The forelimbs of all quadrupeds (frogs, reptiles, most mammals), birds, and humans all contain an upper arm bone (the humerus), two forearm bones (radius and ulna), a collection of wrist bones, and generally five digits. Even when the digits are not seen, the skeletal vestiges are. A horse stands essentially on its middle finger, with the other fingers strongly reduced. A bird's wing consists primarily of its upper and forelimb, while a bat's wing is mostly its fingers. Today we know that even lobe-finned fish have in their lobes bones that can be related to those of amphibia, and the 19th C anatomists understood that the bones of the skull of a fish could be related to those of the quadrupeds, as could the internal organs. As Cuvier argued, and his mentor Geoffroy Saint-Hilaire summed in an aphorism, "There is only one animal".

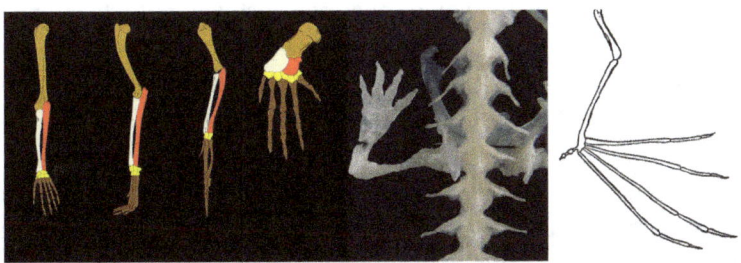

Figure 5.10. From left to right: Forelimbs[xx] of a human, dog, bird, and seal. Tan: humerus; yellow: radius; red: ulna. Note that they are all based on the same architecture. Photograph is the mudpuppy, an aquatic salamander. Drawing is the wing of a bat, showing same construction.

The question was, why should this be? It was a question of sufficient philosophical interest that people like Wolfgang Goethe, Etienne Balzac, and Georges Sand followed such issues with considerable interest. Goethe, in fact, was sufficiently interested that he is credited with recognizing that flowers were modified leaves of plants, earning him a place in the history of botany, and, in the midst of the French revolution of 1830 he was far more excited to attend a debate between Cuvier and Geoffroy than to hear news of the war. Balzac mentioned the debate in his introduction to *The Divine Comedy*.

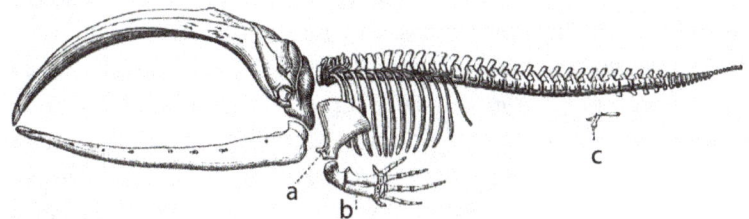

Figure 5.11. Remnants of pelvis and limb[xxi] (c) in a whale. Note also the 5-fingered skeleton of a hand, complete with digits, embedded within the front flipper.

The debate was about God's plan as expressed in this uniformity, which must have derived from a basic wisdom or intended gift of God. Saint-Hilaire, arguing from an almost Aristotelian view, felt that there were theoretical ideal types, such as dogs, cats, etc. for which each species represented modest variations (wolves, foxes). Cuvier felt that there was an original basic type, which God had modified to suit each animal's needs, whether to swim, fly, run, pounce, walk, hop, burrow, or crawl. The frame for the debate, therefore, was not about the potential relationship of the animals but about God's plan for the earth and how God worked.

In England, the arguments were similar. Richard Owen, "the British Cuvier," argued in 1848, "The recognition of an ideal Exemplar for the vertebrate animals proves that the knowledge of such a being as man must have existed before man appeared. For the Divine mind which planned the Archetype also foreknew all its modifications." English anatomists were concerned with the patently absurd issues of anatomy. Walruses and some whales had tiny and completely useless pelvic bones (Fig. 31), which we would today call vestigial, but which had no reason for existing in seagoing mammals. They realized that the bones of a mammal's skull were separate before birth and fused afterwards, and considered this to be a manifestation of God's wisdom: the ability of the bones to slip allowed the baby's head to mold more easily to the shape of the birth canal, reducing damage to both mother and baby though modestly increasing the vulnerability of the baby until the bones fuse. But the bones of a bird's skull were likewise not fused before hatching. Since chipping through the eggshell does not require the head to be molded, the failure of fusion of the skull bones of a bird made no sense. Likewise, it all was very confusing. If God is omnipotent, why does He need a basic design plan? Why not just design a human as desired and a wolf as desired? Why jury-rig a primitive model, or have a practice version?

Improved technology also permitted the examination of embryos. Microscopes and hand lenses were better and, while embryos are typically very tiny, watery, mushy, and nearly transparent, techniques to preserve and dissect them were being developed. Dissecting tools consisted of very finely drawn out or broken glass. The German dye chemists were also contributing. The appearance of pastel colors in dresses derived from the discovery of several dyes, and the microscopists appreciated that what could color wool (a protein) would also stain skin or muscle (primarily protein) and what could color cotton or linen (carbohydrates) would also stain carbohydrates in tissues. There were many surprises to be found in embryos. By 1828, Ernst von Baer had realized that embryonic humans had tails like other embryos and that all vertebrate embryos formed gills. The human coccyx is the remnant of the tail[xxii]; the tail grows so slowly that it is quickly buried in the buttocks, the bones fuse, and it never develops the musculature that would allow it to move. What starts out, in land animals, as a structure very similar to the developing gills of a fish, gradually gets internalized into structures of the throat. The peculiar way in which the aorta branches to deliver blood to the head and forearms—it works, but one could easily come up with other designs, which might be more efficient and less likely to get into trouble—strongly resembles the circulatory system that serves the gills of the fish. By tracing the development step by step, von Baer recognized that some adult structures differentiated from very different starting materials and that even human embryos at some point looked like the embryonic stages of aquatic and tailed creatures, but he had no explanation for the relationship. Several years later, when Darwin had placed the theory of evolution on a sound footing, Ernst Haeckel would make the connection. Although Haeckel's "ontogeny recapitulates phylogeny" (the developmental stages recapitulate the evolutionary stages) was at best an oversimplification, aided in part by perhaps intentionally favorable drawings, today we recognize that evolution can only build on what is present and thus modern forms are built on the scaffolding of the earlier developmental stages of older forms.

Lamarck was more complex. Today we remember Jean-Baptiste Lamarck as the pre-evolutionist who proposed that animals changed to adapt to their needs. Giraffes kept stretching for higher leaves on trees, and so generation by generation their necks grew longer. It was not an unreasonable argument. We can make our muscles grow by exercising, and the muscles atrophy when we do not use them. Plants turn to the sun, and send their roots into the soil. However, by the 20th C, his basic premise was disproven. The renowned German embryologist August Weismann, discussed later because his identification of germ cells (Fig. 32) salvaged Darwin's hypothesis from a potentially fatal flaw in logic, cut the tails from 19 successive generations of mice and showed that, contrary to most opinion, the newborn still had full tails. Thus, he successfully argued, the (absent) tails of the parental mice could not have contributed to whatever built the tails of the next generation. We also know, as neither Lamarck nor Darwin knew, that the information for building a mouse's tail does not come from the tail itself, but from germ cells, which have nothing to do with the tail itself. For Lamarck, a parent who had lost a limb would not have been able to pass the information for building a limb to his or her child. Though knowledge of wars had made that argument ridiculous, in a more subtle manner the argument still had appeal because there was no mechanism to refute it.

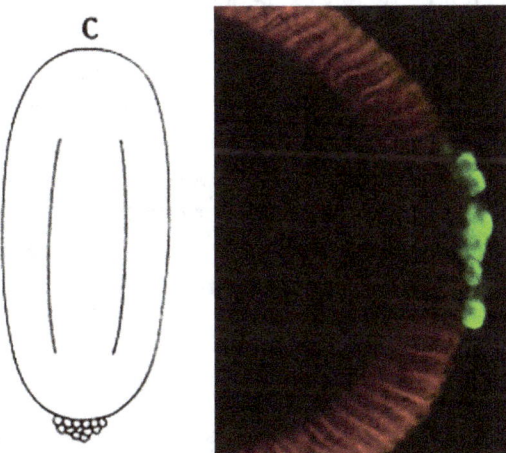

Fig. 5.12. Pole cells as illustrated in the potato beetle[xxiii] (1909) and, by fluorescence microscopy, in Drosophila[xxiv]. During embryonic development, these cells migrate along the gut and ultimately settle in the testes or ovaries to become the future sperm or eggs, or germ cells. The same thing happens in mammals but is much less visible.

Because we tend to treat arguments as adversarial with heroes and anti-heroes (fools or enemies), previously many scientists and even today many laity tend to ridicule Lamarck, but they miss the important issue: He allowed animals to change. He saw, as we all do, that animals were beautifully adapted to function: predators were fast or sneaky, with powerful claws and teeth; herbivores could acquire and chew their plants; moles were equipped with strong paws and claws for digging and tiny ears and eyes that would not be damaged in burrows; animals that lived in cold climates had heavy fur; and animals like otters and seals had forms that were very efficient in water. He also, like Cuvier, studied fossil mollusks which, because of their shells, left abundant fossil records. Like Cuvier, he recognized that the most ancient (deepest) fossils differed most from today's equivalent and that as one got closer to the surface of the fossil bed the resemblance to today's fossils increased. Unlike Cuvier, he considered the change to be gradual—in his terms as the species gradually became better and better adapted to its environment. His mechanism was wrong, but he recognized that species could change.

~~~~~

Geographic distribution

Among the puzzles that the opening of the world gave to those who thought about the issues was the peculiar, unjust, and apparently inexplicable distribution of creatures on the planet. Unlike the issues of anatomical homology, fossils, and vestigial structures, at the time such problems were not, strictly speaking, evidence for the theories of descent with modification (evolution) or natural selection (the primary mechanism for evolution), but they were certainly problems that could not be readily explained by the extant hypothesis, that of special creation and common distribution (from Noah's Ark) and therefore raised a challenge to that hypothesis. By the 18th C, these accumulated challenges were sufficiently compelling to cause intellectuals to mull over the meaning. They could not compare with what was to come, the 20th C analysis of the chemistry and biology of life. These analyses led to such powerful, convergent, and consistent evidence for descent with modification that it becomes impossible to deny. As the great geneticist Theodosius Dobzhansky once said, "Nothing in Biology Makes Sense Except in the Light of Evolution," an argument that, if anything, is more valid today than it was in 1973.

Protein composition

The first issue that came to be of importance was the similarity of proteins. Proteins are extremely large, complex, and fragile molecules but, beginning in the 1950's, researchers realized that antibodies could quite effectively distinguish among very similar proteins, and methods were developed for separating and purifying proteins using their different chemical and physical features. As might have been guessed, proteins that were more similar to each other had similar properties and interacted with more of the same antibodies—cross-reacted, to use the common term. In other words, if you were allergic to the white of a chicken egg, you might be also allergic to the white of a quail egg, but were less likely to be allergic to that of a duck egg; and you were not likely to be allergic to the white of a reptilian egg or any protein with similar properties from a mammal. This allergy could be followed in the laboratory by mixing a sample of your blood with the tested material. From studies of antibodies produced by injecting various proteins into test animals, it became apparent that there were variants among human red blood cells, which we today know as Rhesus (Rh) factors and the major blood types. By antibody testing, these variants were virtually identical in chimpanzees and very similar in other primates. (The term "Rh factor" refers to cross-reactivity with proteins in the blood of rhesus monkeys.) The reaction was much less pronounced against the blood of other mammals and, in decreasing order, with that of birds, reptiles, amphibians, and fish. These similarities and their order proved relatively constant no matter what protein was examined. The similarity extended and was identifiable even to the level of starfish. Once it became possible to purify insulin from farm animals, insulin became the lifesaver it now is, but eventually some diabetics developed immune responses to insulin from cattle, while fewer developed responses to insulin from pigs.

With the advent of the Sanger sequencing technology,[xxv] it became possible to determine, in linear order, the sequence of amino acids that constitutes a protein. It was quickly determined that human, cattle, and pig insulins were very similar, with pig being a closer match than cattle insulin. In fact, by sequence as well as by antibody interaction, the more similar that animals were, the more similar their proteins were. One could build trees of relationships based on the sequences. If, for instance, reptilian hemoglobin differed from frog hemoglobin in that the reptiles had a chain of five amino acids inserted into the middle of the sequence, it was likely that birds and mammals would also have that sequence, suggesting that the birds and mammals were more closely related to the reptiles than to the frogs. These trees of relationships coincided almost completely with the trees of similarity constructed on the basis of anatomy or apparent fossil record.

Nucleic acid composition

As dramatic as the protein similarities are, nothing tops the similarities determined as molecular biologists got their first look at genes. To anticipate where this argument is going, if we were to encounter our first Martian, and its native language would prove to be (to follow the conceits of certain members of the human race) English, spoken in the accent of a 21st C evangelical preacher from Texas, we would have to conclude that there was some fundamental connection in the universe. In the same sense, the genetic code is universal. The first hint was relatively crude but intriguing. If you heat DNA, the strands unwind. If you let it cool, slowly, the opposite strands find each other (basically, they keep bumping into other strands until they bump into one into which they fit or bind tightly, and then they stay). The physical properties of double-stranded and single-stranded DNA differ, and one can therefore detect these associations. If you mix the DNAs of two animals that are only modestly related, the strands may fit into each other, but badly. They will be less stable and come apart at lower temperature. Thus you can tell how similar one DNA is to another by the temperature at which a hybrid strand becomes unstable. If you do this with many animals of differing degrees of relatedness, say our humans-apes- other mammals-birds-reptiles-amphibians-fish series, you find that the similarities of their DNAs follow closely the similarities adduced by all other criteria. Today, if you sequence these DNAs, you find that the hierarchy is valid even down to the level of sequences. Even more astonishing, and more germane in terms of the language analogy, were the findings that the mechanisms of protein synthesis, the genetic code, and the structure of genes argued for lineal descent[xxvi].

~~~~~

Chapter 6: Natural Selection

Most people have only the vaguest idea of what Darwin actually said. There are at least four elements to "Darwinism". The **first** is the existence of previously existing species unlike anything that we see today. The **second** is the possibility that these previously existing species were ancestors to the species that exist today. After all, the title of Darwin's most famous book is "*Origin of the Species*". The **third** is the possibility that the earth is much older than biblical chronology, and therefore that there has been enough time available for one species to change into another. The fourth is the **mechanism** by which this change could come apart, what Darwin calls Natural Selection.

The first three topics are those that are often challenged by Fundamentalist Christians, although they are frequently misunderstood and challenged for the wrong reasons.

Regarding the first point, there is no doubt that fossils exist and that they are different from modern forms of life. Since the 18th century, they have been recognized as existing in layers of the earth that are deeper and for which therefore materials have accumulated above them. They therefore are quite a bit older than modern forms. What has been disputed is how old they are, and what relationship they have to modern forms. For instance, interpretations ranged from efforts by Satan to confuse humans and to tempt them to impious thoughts to practice versions of modern animals made by God. The second point was resolved by the recognition that the deeper one went into the earth, the older the samples were, and that the earlier fossils differed more than the later fossils from modern forms. That left the big question of whether there was enough time for all of this to occur. The approximately 6000 years attributed to the age of the Earth by the Bible would not be nearly enough. From the beginning of the 19th century on, this issue was being resolved: Many ways of measuring the age of the earth were being devised, and they all led to the conclusion that the earth was much older than 6000 years. If one were to deny one form of measurement, then one would have to deny that complete line of reasoning and therefore that science, or one would have to invalidate all the other measurements as well. Thus, for anyone who was willing to admit the possibility that the Bible might not be the final arbiter of Natural Law, the three issues attributed to Darwinism were already resolved. All science is built on three pillars: Evidence, Logic, and Falsification, with Hypotheses coming in to play somewhere early in the process. By the mid-19th century, the evidence was all there. Fossils were real, they changed over time, becoming more and more like modern versions, and there had been plenty of time over which this could occur. What was lacking was a mechanism by which this could occur – the logic of Darwin's hypothesis. His hypothesis was that evolution had occurred, and that new species could be generated, by the process or mechanism of Natural Selection.

The hypothesis of Natural Selection is based on several observations and inferences. The first observation, relatively obvious, is that all creatures produce more young than is necessary to duplicate the current generation. Humans, estimated by the expansion of populations in virgin territory such as when they first settled the Falkland (Malvinas) Islands, can average eight children per couple. Fish can lay 10,000 eggs per year, and trees produce 50,000 seeds per year or, over a hundred-year lifespan, 5,000,000 seeds. In a five-month breeding season, a single pair of mosquitoes could produce one nonillion (10^{30} or 10 followed by 30 zeros) descendants. Even a pair of birds or mammals that produces only one young per year will, over the reproductive lifetime of the pair, produce far more young than one replacement pair. If the population does not grow like wildfire—by and large, there will be as many fish in the stream next year as there are this year—then something must limit or kill off this tidal wave of young before they mature. Therefore the first inference based on observation (that only a fraction of all offspring survive) was one that could have been made by any hunter, fisherman, or farmer from at least the time of Aristotle:

- Inference: All the young must struggle to survive, and it is likely that they compete with each other. There could be any number of limiting factors: food, water, suitable hiding places or suitable locations for nests, and only a few individuals get the good opportunities.
- The second inference comes from observations that, again, any person in contact with other animals or plants could have seen, but required a different philosophy to address: Inference: Among all species, individuals vary in many characteristics. This is especially true for domesticated species, including ourselves (where, we know today, the usual operators of natural selection, predators, do not limit the size and color variants), but it is true for all species. Even penguins can identify their partners, and penguin chicks can identify their parents.
- The change in philosophy was the movement from the idea that lasted through Linnaeus, that each species was its private cloister, separated by distinct characteristics from similar species. Linnaeus' effort was so thorough that it eventually undermined the whole principle; but it was Darwin and Wallace who embraced that variation as a natural and important force in determining the shape of the biological world.

Figure 6.1. Selection pressure. Yellowtail snappers, barracuda, or other predators swim into a school of smaller fish, which quickly scatter, leaving a clear area around the predator. The predatory fish quickly feints an attack, scattering the fish in a new direction. Eventually one fish fails to keep up with the others and is left in the cleared area, where it is quickly consumed. For a better sense of this, see higher resolution figure.

The next inference again comes from an obvious observation, one which all farmers, hunters, fishermen, and parents (but not scientists) made and could have extrapolated, had they appreciated the implications:

- Inference: Much of the variation among individuals is heritable, or passed from one generation to another.

Children normally look like a combination of their parents. If you prefer Dalmatian dogs with few spots (show dog criteria) then you breed two Dalmatians (of course; not other breeds of dogs) that have few spots, rather than large black patches. If you want guppies with brighter colors, then you pair the male and female with the brightest colors. If you want a faster race horse, you choose as parents the fastest horses you can find. Of course, other factors may come into play. A peach tree that gives very sweet peaches may do so because it has the best soil and sun, not because it has a particular inheritable characteristic. But often the individual with the better characteristics will be able to pass the characteristics to the next generation. One of Darwin's great leaps of intuition was that he recognized the parallel between what breeders do and what Nature does. Breeders, say, dog breeders, have over centuries changed dogs from creatures very similar to wolves to all the wildly different forms we have today by selecting which dogs to breed (the definition of domestication). In other words, they decided which characteristics to pass to the next generation, excluding those that did not fit their goals. Darwin recognized that Nature could, in an impersonal and undirected fashion, do the same thing. (Wallace did not see the analogy and assumed that the variation between different populations, rather than individuals, was important, but he came to the same conclusion.)

- Inference: Which variants survive is not a random choice. The individuals that leave descendants to the next generation will be those that function best in their environment.

Figure 6.2. Camouflage. Left: Lizard, South Africa; Center: Cheetah, South Africa; Right: Flounder, Mexico.

To some extent, survival will be a matter of chance. Of the thousands of seeds that a tree produces, many will be blown or transported to very inhospitable terrain, landing on rocks or in the water, where they cannot grow. Squirrels, birds, or other mammals may eat many of the seeds. Some seedlings will be destroyed by fire, desiccation, or frost. But sometimes, maybe often, survival will be because of some inheritable characteristic the Malthusian recognition of the fact that all creatures breed far more young than can be supported by the local conditions, and that many of these young do not survive.

Figure 6.3: Adaptations for camouflage. Sea horse[1].

Darwin pointed out that all species vary considerably between one individual and another, a conclusion that derived from Linnaeus's effort to try to classify every species that existed on Earth. Darwin argued that our inability to be certain of our classification of each individual into one species or another was not a flaw in our ability but the mechanism by which evolution could occur. He argued that, as breeders of plants and animals select individuals showing certain characteristics as the ones that they will use for further breeding, and that they thus alter the species in the direction that they prefer, Nature would do the same thing by selecting those animals most capable of leaving young to the next generation and thus gradually altering species in a natural fashion. He argued that some individuals would be more fit than others. By "fit" he did not mean bigger, stronger, more handsome, or necessarily any other positive feature. He meant simply, the ability to survive and leave young to the next generation. Thus there would be a process of selection. Some individuals would, by virtue of their variation, the more likely to survive and leave young. It might be strength or speed. A gazelle might be able to run faster and therefore outrun a charging lion. However, it might also be better camouflaged (Figs. 6.1 and 6.2), or much more skittish, and therefore diverge from the herd as it ran. It might even, like a cockroach, be smaller and more capable of slipping into a tiny area where the lion might not be able to attack. It might even mean giving up a free, mobile life for a safe, hidden cocoon (Fig. 6.3).

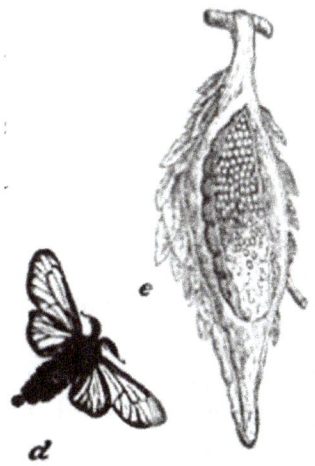

Figure 6.4. A bizarre lifestyle choice. A male bagworm (left) is a clear-winged but otherwise conventional moth. The female never develops wings or legs, but remains a sort of furred worm inside her bag. She emits a scent by which the male finds her, and she mates, lays her eggs, and dies within the bag. From The Moth Book, by W.J. Holland[ii]

This then is what natural selection means: Individuals within a species will vary. Much of this variation will be heritable. Some of the variants for whatever reason will be more likely to survive and produce children then other variants. Thus the new generation will have more individuals with the characteristics favoring their survival. Over time, these species will gradually change to resemble those individuals more likely to survive. This process over many generations will ultimately form a new species – the "origin of the species". Darwin argued that this process could be nature's version of what obviously was successful and effective when breeders of plants or animals chose what they wanted in domestic creatures. Wallace did not see the parallelism between domestication and domestic breeding and natural selection, but he nevertheless came to the same conclusion[iii]

~~~~

Chapter 7: The Rediscovery of Mendel

One of the curious things about science, or indeed, many human endeavors, is coincidence. It is often pointed out to students that different scientists almost simultaneously come to the same discovery. Commonly cited examples are Tesla and Edison, building electrical networks for countries; Salk and Sabin, producing the first polio vaccines; Darwin and Wallace, simultaneously positing a theory of natural selection; and three separate laboratories (those of Hugo de[i] Vries Carl Correns[ii] and Erich von Tschermak[iii]) rediscovering Mendel in 1900. These are often explained with a sort of wonderment, suggesting how remarkable it was that such an idea could lie fallow for immeasurable eons of human existence and then suddenly pop into two or more heads.

The reality is somewhat different. As we describe in discussing the polio vaccine[iv], the age of exploration, and the relationship of Darwin and Wallace, ideas congeal around intellectual ferment and discoveries of the time. Exploration was driven by search for resources and required the development of specific tools, while in turn the explorers encountered situations and creatures that caused them to doubt the stability posited in classical and religious texts. Darwin and Wallace had available to them the confusion of the variations within species, knowledge of the peculiar distribution of organisms around the world, an understanding that the earth could be much older than was previously supposed, a sense of the historical sequence of fossils, and Malthus' recognition of a struggle for existence. Developing the theory of natural selection was a matter of connecting the dots. Similarly, neither Sabin nor Salk could have developed a vaccine had it not been possible to grow poliovirus in the laboratory, a technique that had recently been achieved by John Enders. The coincidence really comes from the fact that the same questions are in many peoples' minds. At some point paths to resolving these questions become visible, and those who are ultimately touted as the best scientists who see, design, or invent that path by a leap of the imagination or an ingenious trick.

Sometimes the answer hides in plain sight but is ignored because either it does not make sense at the time, everyone is convinced that the system works in another fashion and even the best scientists are not open to alternatives, or the answer is too boring, mathematical, or otherwise unexciting that no one pays attention. Such was the case with Mendel, whose paper, though published in a well-read journal, lay unremarked for 34 years, only to be rediscovered more-or-less simultaneously by three laboratories. It provides an interesting image of science to realize why this happened. The story includes boring papers, the intellectual fixation (we could say 'stupidity') of thinkers in ignoring the obvious, and the force that ultimately inconvenient facts and logic play in coercing self-correction.

Darwin was typically so thorough and so timorous about suggesting that the biological world could develop without a grand master-plan (devised by God) that he literally overwhelmed his readership with detail, documentation, and argument. Modern readers of the unabridged version of *Origin of the Species* may find it surprisingly tedious. This is because, imbued today as we are with the logic and evidence of his argument, we are convinced after three examples that he is correct—but he goes on to add another seventeen examples. However, for the mid-19th C, his massive assault on the older ideas was well appreciated and convinced many previously skeptical readers. Those who followed his argument quickly conceded that he was most likely correct.

However, once the ideas of natural selection and descent with modification (production of a new species from an ancestor) had sunk in, the next level of question: how does it work? became apparent, and here was a problem. According to common understanding, inheritance worked by some sort of distillation process. Traits from both parents (sometimes, according to prejudice, only the father; though the mother could be abandoned or mistreated for not producing sons) were delivered, most likely through the blood, to the semen,* whence they were delivered to the womb. For those who accepted a role for the mother other than "flower pot for the baby" her traits likewise found their way to the womb, the human egg or oocyte not yet having been identified or understood. (It had taken some time for microscopists to realize that sperm were part of the fertilization process, not some bacterial contaminant of semen.)

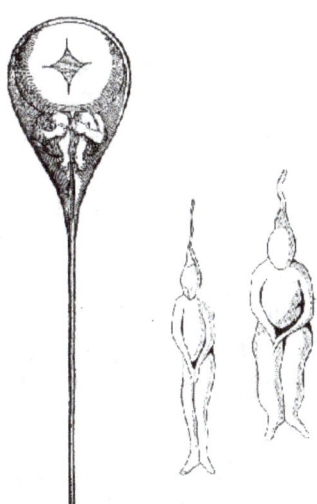

Figure 7.1. What microscopists saw as little people embedded in sperm. In reality, what they could see was at the limit of the resolution of their microscopes, and optical effects could lead to the illusion of structure within the sperm head. v

This idea has consequences. For instance, an amputee has nothing to distill for an arm or a leg. Other modifications throughout life would likely have consequences.

A muscular laborer should be able to provide more "muscle stuff" while someone whose skin was severely disfigured by smallpox might provide only poor quality "skin stuff". This was the basis of Lamarck's defense of the proposition that acquired characteristics could be inherited. The idea seemed reasonable. Strong laborers tend to have strong children; organs such as muscles grow with use and atrophy if they are not used.

A fundamental, perhaps fatal, consequence for the premises of natural selection would be that any trait that was potentially worth selecting for would unfortunately be diluted out of existence within a few generations. Let's take an example: I bear a new mutation that gives me a huge selective advantage over my rivals. Let's suppose that I can photosynthesize all of my food, and that therefore I don't have to hunt or expend any energy in acquiring food. I can devote all my time to other worthwhile tasks, such as eliminating my rivals or procreating. However, the partner I choose does not have this mutation: the mutation is new and unique to me. Therefore, assuming that the woman contributes to the baby (reasonable since children often look like their mothers) then our child will have only half the ability for photosynthesis, and our grandchildren, presuming no incestuous relationships, will have only one quarter of my ability. Continuing down the line, it is obvious that the marvelous trait will dilute to insignificance before any selective advantage could be achieved. By the late 19th C, this issue was becoming a major stumbling block to defending Darwin's hypothesis. After all, science requires evidence AND logic, and the logic was failing.

Of course, in hindsight this line of reasoning was totally ridiculous, as almost any peasant or animal or plant worker could have pointed out. Particularly in northern or western Europe, where red hair is more common, most people could have pointed out that Jane had Grandpa Pete's flaming red hair, even though neither Grandpa's daughter, Jane's mother, nor Jane's father, had red hair. Likewise various animal colors or markings could show up sporadically. Thus there were two conclusions: 1: traits such as colors were NOT diluted from generation to generation, but could resurface uncontaminated in a later generation; and 2: a trait such as a color could remain hidden in what we can now call a "carrier" without manifesting the slightest indication of its existence. But these tidbits from the real world did not effectively penetrate the ivory towers.

Except for Mendel. He, like many others, was trying to understand how inheritance worked, but he asked his questions in what proved to be an infinitely more practical way. Though he worked with plants, his questions could be summed as follows: Instead of asking, "Does this child look more like its mother or its father?" he broke the questions into smaller components such as, "Does this child have the hair color of its mother or its father?" "Its eye color is that of which parent?" "Which parent gave the shape of the child's lips?" "Of its ears?" et c. For the peas with which he worked, he examined, independently, height of the plant; whether it produced purple or white flowers; whether it produced green or yellow peas; and whether the mature peas were wrinkled or well filled out and round. When he did this he got quite consistent results:

Parental (P)		First Filial (F_1)
Tall plants x short plants	→	all tall plants
Plants with purple flowers x plants with white flowers	→	all plants with purple flowers
Plants with green peas x plants with yellow peas	→	all plants with yellow peas
Plants with wrinkled peas x plants with round peas	→	all plants with round peas

Fig. 7.2. The result of Mendel's first-level crosses. The parental generation consisted of pure-breeding plants crossed with pure-breeding plants that showed different characteristics. The seeds produced the plants of the first filial, or F_1, generation.

In each case one trait disappeared entirely, to be replaced by the other trait. There appeared to be no dilution. (See below to see how this works.) He called the trait that stayed visible Dominant (in the original German, Domineering), over the recessive trait that was no longer seen.

He then continued the experiment, using the plants that resulted from the cross, the children or first filial generation, and interbred them (crossed them with their brothers and sisters. Since most plants are bisexual, incest doesn't really count. Most plants have a means of discouraging self-fertilization, like having the male parts and female parts mature at different times, but in domesticated plants this can be overcome.) The result, the grandchildren or second filial generation, was quite surprising:

First Filial (F_1)		Second Filial (F_2)
all tall plants	→	¾ tall plants, ¼ short plants
all plants with purple flowers	→	¾ purple flowers, ¼ white flowers
all plants with yellow peas	→	¾ yellow peas, ¼ green peas
all plants with round peas	→	¾ round peas, ¼ wrinkled peas

Fig. 7.3 The result of Mendel's second-level crosses. The F1 generation plants were crossed among themselves. The seeds produced the plants of the second filial, or F_2, generation.

He also did double crosses, those involving two separate traits. For instance, if he crossed tall plants with purple flowers to short plants with white flowers, in the F_1 he got all tall plants with purple flowers. If he then crossed the F_1 plants, he got all combinations, but in a typical ratio: approximately 9 tall, purple flowers; approximately 3 each of short, purple flower and tall, white flower; and 1 short, white flower plant.

These results carried several important points and a couple of tidbits that might have proved a bit embarrassing. The important points were that some traits, here the recessive traits short, white, green, and wrinkled, could reappear intact and uncontaminated after existing hidden behind the dominant trait. They were not diluted at all. Second, he worked out that this distribution in the F_2 would be what one would expect if each trait existed in two copies and all traits—each copy of each trait— appeared in each individual plant entirely independently of each other copy of the other trait. For instance, for the 3:1 ratio, you can imagine flipping many times a pair of quarters. One quarter of the time, you would get two heads, one quarter heads, then tails; one quarter, tails, then heads; and one quarter, two tails. If you count only the times that you get at least one head, then ¾ of the time you will get at least one head. (That too makes sense, but wait for another paragraph or two.)

The embarrassing points involve a couple of issues. The first was more amusing than embarrassing. What Mendel was exploring was whether his model followed the laws of random distribution that were being worked out in the emerging science of statistics. This latter, the science of statistics, was being developed for the benefit of the growing casinos of the 19th C, the owners of which of course wanted to peg winnings of clients at below the rate of random return. There was nothing unethical about interest in this type of mathematics, though presumably, since Mendel was a monk, his interest was purely intellectual. The second point is a bit more dubious. Several scientists have tried to repeat Mendel's experiments and, while they always get approximately 3:1 or 9:3:3:1 ratios, the ratios are never as close as Mendel reported. Most likely, he made judgment calls—whether a yellowish-green pea should be counted as "green" or "yellow". Once he saw the pattern developing, those judgment calls inadvertently favored the ratio he was looking for. If he was a little short of green pea plants, the yellowish-green pea plant would be green. This is not to say that Mendel was dishonest. It is one of the arguments we present to students about the importance of very careful controls and blind studies. The final question is why his paper was ignored for so long, and why it was rediscovered. First, Mendel was interested in the question of whether traits could distribute randomly. The ferment over evolution had not yet sunk in, and he made no effort to suggest that it would resolve an important question in evolution. Unlike the famous paper by Watson and Crick[vi] describing the structure of DNA, in which they recognized and called attention to the fact that the structure itself could explain how genetic information could be passed from generation to generation and therefore made it extremely likely if not necessary that DNA be the genetic material itself, Mendel did not state or emphasize that his model could resolve the grand conundrum of evolutionary theory. Indeed, the conundrum had not been articulated at the time he published his paper in 1866. Without this emphasis—the second point—the paper was a relatively dry, abstract, mathematical paper, a model of a relationship, of the kind that many biologists pass by. In the same sense that there are elegant equations to calculate the trajectory of a football but there are shorthand approximations and estimates that give pretty good results, most scientists don't bother with the equations to establish relationships. Besides, the paper was on peas rather than on animals, and thus was dismissed as belonging to the world of agriculture.

By the end of the 19th C, the dilution problem had become a serious impediment to the logic of the theory, for which the evidence was becoming indisputable, and it was becoming more and more urgent to prove the dilution problem wrong or to find another means of explaining evolution. Thus the three researchers mentioned above were all working on the problem, began to find similar solutions, and rediscovered the now relevant paper by Mendel. If traits were not diluted, then the theory of natural selection would work.

One final point: the meaning of dominance and recessiveness. Generally, a trait is produced by a gene, that is, the information encoded in DNA to make an enzyme (a protein catalyst) that can carry out a specific reaction. In most organisms each individual gets two copies of each gene, one from the mother and one from the father. The purple flower exists in a plant that has the enzyme to make the purple pigment. The white flower is formed if that enzyme is lacking. If either copy of the gene is intact, the pigment can be made, and the flower will be purple. (In fact, there might be a little less pigment than when two intact copies are present, but the difference is generally not noticeable.) So, the purebred purple-flowered plant has two copies of the purple gene, the F_1 plant has one copy but is still purple, and the white flowered plant has no good copies of the purple gene. The F_1 plant can give, randomly, either the good purple gene or the bad purple (white) gene. Each F_1 can do the same, so that the possible combinations are, like our quarters above, one can get purple, purple (appears purple); purple, white (appears purple); white, purple (appears purple), or white, white (appears white)—our 3:1 ratio. This point is illustrated in an animation that demonstrates how chance leads to these results[vii].

You may be aware that, in plants, yellow is often the absence of green. Tree leaves turn yellow and red in the fall when green chlorophyll is withdrawn, revealing the pigments that were always there but hidden by the overpowering green. So why is the yellow pea dominant rather than recessive (absence of chlorophyll)? It turns out that the yellow pea contains an enzyme that degrades the chlorophyll in the pea, leaving the yellow[viii]. The green pea cannot make this enzyme and therefore retains the green color. Again, the recessive trait is due to the lack of an enzyme.[ix]

Chapter 8: The history of our planet

We can begin to understand the sequence that led to the proliferation of species, but this does not give us much sense of how we got species in the first place. This is the first of several questions that we have yet to address. These questions are: Where did species (living organisms) come from; why did it take so long to get to the world that we know; what mechanisms drive the process; where do we (humans) come from; and what does the entire story mean? With a bit of common sense, to which we can add the assumption that physical laws do not change with time or from planet to planet, together with some knowledge of these physical and chemical laws, we can generate a fairly detailed image of the physical nature of our early planet. To make our story as minimally boring and complex as possible, minerals and crystals form differently depending on the temperature and pressure under which they form. The element carbon can under different circumstances form diamond, graphite, or coal. Iron will rust in the presence of oxygen; we paint iron objects to prevent oxygen from reaching the iron. Ice sliding along a surface, as in the movement of glaciers, will form patterns different from those caused by flowing water, thus distinguishing terrains formed or marked by temperatures above or below freezing. There are even intermediates. Many crystals can incorporate varying amounts of water, oxygen or salts, producing slightly different forms and different minerals in each condition. From such information we can we can read that, at the beginning of the earth, estimated by radioisotopes to be approximately 4.5 billion years ago, the earth was quite hot and it lacked oxygen. This would be consistent with the argument that the earth had separated from the sun and that oxygen, an extremely reactive molecule, would have reacted with anything available—for instance hydrogen, forming dihydrogen oxide or water, steam, or ice, and ferric oxide, or rust—leaving no oxygen free in the atmosphere. The planet was too hot for water to exist. Making two very reasonable assumptions, we can conclude that life did not exist.

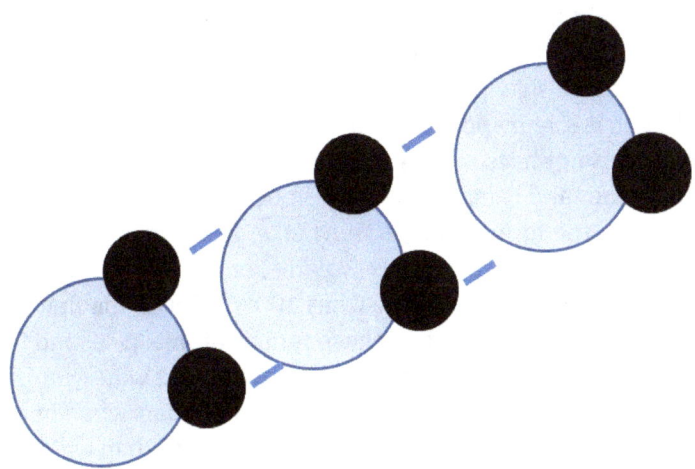

Figure 8.1: Water. Because the hydrogens (black) are aligned to one side of the oxygen (lighter color) each molecule has a slight negative charge toward the oxygen end and a slight positive charge toward the hydrogen end. Since positives and negatives attract, The molecules tend to cling to each other and align. The alignment creates the hexagonal patterns of snowflakes, since the angles of the hydrogens are approximately that separation; and the adherence of the molecules makes water much less volatile than methane, which is approximately the same molecular mass.

Requirements for life

The assumptions concern our understanding of the basic requirements for life. Most scientists are comfortable with the idea that life cannot exist in the absence of complex carbon compounds and water. This is the reason that interplanetary probes search for these carbon compounds and that newspapers use headlines to announce evidence for water on the planets of our solar system. But why should this be so? Let's first look at water. Water is a very unusual molecule. It consists of two atoms of hydrogen and one of oxygen, which add up to a molecular weight of 18 meaning that, if water were a gas at freezing temperature and one atmosphere of pressure, a standard volume of that gas (for historical ant technical reasons about 6 gallons) would weigh 18 g, or about ½ oz. Most molecules of that size are very different. Methane, for instance, weighs almost the same, but it is a gas that becomes a liquid only at -161° C (-258° F). The reason for this is that water molecules tend to stick to each other. In methane, 4 hydrogen atoms are more or less symmetrically attached to one carbon atom. The molecule has no net charge, and the molecules do not attract or repel each other. They ignore each other, and a molecule this small has enough energy that they career around randomly, forming a gas. Other small molecules do the same: oxygen (two oxygen atoms linked together, molecular weight 32); carbon dioxide (one carbon plus two oxygens, molecular weight 44); acetylene (two carbons plus two hydrogens, molecular weight 26). In fact, you can see this tendency by comparing similar molecules of the same series. Molecules similar to methane but containing 2, 3, 4, 5, 6, 7, 8...16 carbons are respectively a gas (ethane), a highly volatile liquid (propane), liquids of steadily less volatility (butane, pentane, hexane, heptane, octane) and finally, at about 16 carbon atoms, a waxy solid like paraffin. Water differs because it is polar. The hydrogens are not symmetrically distributed but rather are more to one side. Because of this, like a miniature magnet, the end with the hydrogen is slightly positive and the other end is slightly negative. The molecule itself does not have a charge, but two or more molecules can interact by attracting each other and lining up positive to negative end. This alignment is directly reflected in the formation of a snowflake. Because these molecules stick to each other, water is considerably less volatile, becoming a solid at 0° C and remaining a liquid until temperature reaches 100° C. This is the first important property of water: that it is a liquid at these temperatures. A second feature related to this structure but less directly is the fact that as water cools to freezing, the atoms align in such a fashion that they are farther apart when immobilized in ice than they were in the water. Thus the ice is less dense than water, and the ice floats. Because ice floats, it insulates the water below it and prevents it from freezing. Most liquids do not do this. The solid form sinks into the liquid. If water did this, lakes and rivers might freeze solid in the winter, making it extremely difficult for large, organized living creatures to survive. Life needs chemical reactions to continue, and chemical reactions come to a near stop in solids. That's why we can store foods in freezers almost indefinitely. A third feature of water is that, contrary to popular opinion, water is one of the best solvents that we know. Think about that. We can dissolve almost anything in a water-based medium. Even the fact that we can taste and smell gasoline in water that has come into contact with it means that some of the gasoline has dissolved in the water. For chemical reactions to take place, two or more molecules must physically collide. They cannot do that unless they all can physically move in the same medium, or at least one can move to encounter the other. If they do not dissolve in the medium, they cannot interact. That's why layering a homemade jam with paraffin (or even an oil) will preserve it: it keeps oxygen away. But it is the fact that water is a liquid between 0 and 100° C that is most important, because of the importance of carbon and the peculiarity of carbon-based molecules.

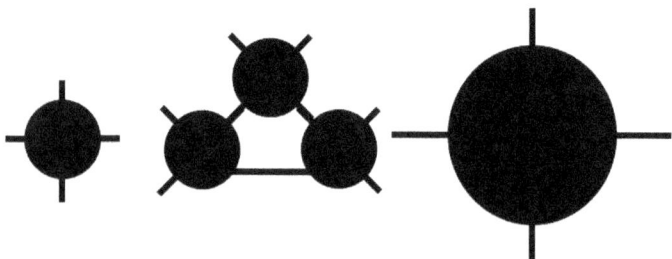

Figure 8.2. Carbon (left) has for bonding electrons, which can bond at various angles, such as in propane (center). Thus carbon can form molecules of many shapes. Similar atoms, such as silicon (right) are too bulky for stretched and bent bonds to easily form.

Carbon is a very special atom, and it may be unique among all elements. It has four sites at which it can bind other atoms, and it may create these bonds by giving, sharing, or taking electrons from other atoms. Importantly, these bonds can bend a bit so that, if we picture the atom as a sphere, the bonds may be at right angles to each other, as in methane, or aligned two by two on opposite sides, as in carbon dioxide. If the bonds are really strained to make their connections, as in the molecule acetylene (a two carbon compound with each carbon having one bond to hydrogen and three to the other carbon) the strain acts very much like a cocked spring, ready to release energy. Thus acetylene is extremely flammable. But the important point is that the carbon bonds can move and be bent. Because of this and because carbon can share electrons, an infinite number of combinations consisting primarily of carbon can be formed, generating molecules as diverse as paraffin, plastics, hair, horns, hemoglobin, egg white, muscle, cartilage, and meat tenderizer. No other atom can do so much. Silicon, for instance, is in many ways similar to carbon, but the core of the atom (the atomic nucleus) is so big that the four bonds cannot interact with each other. Compared to carbon, silicon is like the cartoon character who accidently gets hooked to a balloon- filling tank and blows up like a balloon, his arms and legs hopelessly splayed out into non- functional positions. Silicon, with oxygen, can form long chains that can create silicon rubbers or crystalline structures like sand, but it otherwise provides little adaptability.

So carbon is essential, but it has its problems. As you might guess, if something can make many flexible bonds, many of the bonds may not be very strong. You can glue parts together in many different ways, but unless there is a physical interlocking of the parts, the assembled object can be easily damaged. At worst, carbon and carbon-based molecules can easily interact with oxygen: they can burn. That is the whole point of coal and petroleum. To a lesser extent, most carbon-based molecules are not very stable above 50° C (122° F) and may even show considerable decay above 40° C (104° F). Thus: egg whites cook; our skin burns; meat turns brown (a protein called myoglobin is destroyed); lobsters turn red (a bluish protein called astaxanthin precipitates); and we become delirious (the fats that make up the cell membranes of our nerves begin to soften, making the nerves less stable). Mammals and birds, the Porsches of the animal world, have contrived to run their body machines at near the limit of stability, maximizing the rapidity of chemical reactions. Also, because many of the carbon-based structures are held in shape only by weak interactions with other parts of the molecule, small changes in acidity or salt content can seriously disrupt their structure. Thus milk curdles if bacterial activity produces too much lactic acid, and blood proteins precipitate if the blood is diluted with fresh water. These constraints—the facts that carbon is the only atom adaptable enough to build living creatures, chemical and particularly carbon-based reactions take place in water, and carbon-based molecules are usually not stable above about 50° C— place strong constraints on where we might predict that life can thrive. This is why space probes search for evidence of complex carbon-based molecules, water, and temperatures above freezing but below 60° C.

The early earth

It took the earth about 500,000,000 years to cool to the point that water could condense from steam, at which time rain poured incessantly on the earth, extracting soluble salts from the land and carrying them to the forming seas. The seas accumulated salts to about 0.2%, compared to the approximately 0.8% we see today. The increased salt has come from continual washing out of salts from the land. Because of the physics of how raindrops form and fall, they produce charge differences between the atmosphere and the earth, leading to continuous lightning strikes. Also, unrelated but relevant, the moon was closer to the earth, generating tides hundreds of miles inland. All of these are relevant to our image of how life appeared on the earth. At this time there was no free oxygen, but there was lots of hydrogen sulfide, sulfur dioxide, methane, ammonia, and nitrogen. We can read this from the minerals that formed at the time. For instance, there is no free ferric oxide (iron rust). Rather, iron is complexed with carbon, and nitrogen, forming ferricyanides. These aspects of the early earth were worked out by the early 1950's. Then, in 1953, Stanley Miller and Harold Urey published an answer to a provocative question: could organic molecules form spontaneously in such an atmosphere? By "organic molecules" they meant the relatively small carbon-based molecules that formed the building blocks of giant macromolecules. Amino acids are the building blocks of protein; simple sugars such as glucose, of starches, glycogen, and cellulose; fatty acids, of fat; and nucleotides, of nucleic acids. By "such an atmosphere" they referred to the mixture of gases described above, mixing with hot, agitated water containing some salts, and with constant lightning. In today's atmosphere organic molecules do not spontaneously form. They react readily with oxygen and are destroyed. But in an anoxic world of toxic gases and hot, salty water, with electrical discharge, was it possible? Miller and Urey concocted such a preparation and put it in a sealed flask that they could agitate and through which they could deliver frequent electrical discharges. They let it "cook" for several weeks, after which they opened the flask and analyzed the solution. They found sugars, amino acids, fatty acids, and nucleic acids. In the atmosphere of the early earth, it was possible—and, given the time available—extremely likely that the building blocks of life would be formed. This was a step, but it is a long way from your sugar bowl to a living creature. The next step would be to spontaneously make macromolecules from the building blocks. This is impossible today: living creatures, including bacteria, have enzymes that can rapidly destroy anything that could slowly and accidentally be formed. If bacteria are not present, oxygen will disrupt the molecules. The world tends from order to disorder, and from higher stored energy to lower stored energy. When wood reacts with oxygen and burns, heat is given off as stored energy is released, transforming the highly ordered cellulose into disordered carbon dioxide and water. The same thing happens, albeit much more slowly and therefore with imperceptible heat release, as oxygen destroys other molecules. However, oxygen and bacteria not present, and we have millions of years. Again, based on analysis of long-term Miller-Urey type experiments (we now have much more sensitive means of identifying rare molecules in solutions) we can document that macromolecules can form. In these conditions they do not decay, or decay only extremely slowly. You can see this in any grocery store. Canned foods and (in hiker/climber stores) irradiated foods, which are sterile and are stored without oxygen, can persist for many years, even at room temperature. We cannot prove that macromolecules formed in the early earth, but we know that they could, and there is no intellectual argument to say that they did not. Besides, meteorites and other extraplanetary samples often contain at least simple organic molecules, and sometimes more. These findings provide a major argument for those who believe that life came to us from another planet. (This belief has a certain appeal, but it begs the question of how life arose; it just pushes it off to another location.) Living creatures are not molecules, no matter how big the molecules get. Viruses are collections of molecules, but they are not truly alive. The molecules of viruses can penetrate a cell, but to reproduce, the virus has to capture the cell's machinery and use it to make new virus. In order to be alive, which means to use energy to create higher structure and order from simple starting products and to be able to reproduce one's likeness, you have to control and direct the reactions. The environment can be very variable. To be alive, an organism must be able to control its environment sufficiently to carry out these controlled reactions. One way to do this is to isolate the living from the non-living. Luckily, there is a way to do this that does not require life to set it up.

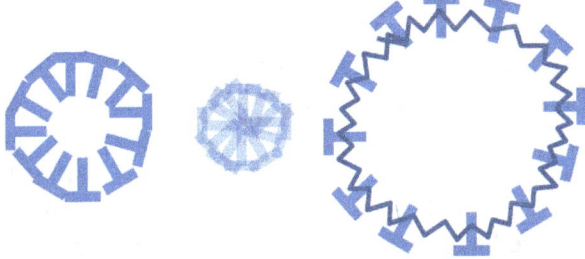

Figure 8.3. Coacervate structure. A coacervate is essentially a bubble of one liquid inside another liquid. Its membrane, consisting of fatty acids, fits comfortably at one size (left). If it is too small (center) the squeezing creates a pressure that allows it to absorb fluid and expand, while if it is too big it is weaker and can split into two. Thus coacervates tend to normalize to a specific size.

Precursors to living creatures

In 1922 the Russian biochemist Alexander Oparin pointed out that simple lipids (fats), such as those that could form spontaneously, had very interesting properties. You vigorously shake salad dressings, spray paints, insecticides, and many other things before you use them. These are mixtures of fats or oils (or something similar) and water. The shaking breaks the fats into tiny droplets, where because of their size they remain suspended long enough for you to distribute this suspension, termed an emulsion, as you wish. If you shake or blend the mixture even more vigorously you can make the droplets so fine that they keep colliding with each other and cannot settle out, so that they remain suspended almost indefinitely. In this state the mixture is called a colloidal suspension. Homogenized milk is such a suspension. Oparin pointed out that there was another possibility. Under certain conditions, the suspension comes out differently. Rather than a suspension of oil droplets in water, the droplets can be little hollow spheres with membranes one or two molecules thick, containing aqueous medium on the inside of the sphere as well as floating in aqueous medium on the outside. In this condition the mixture is called a coacervate. Coacervates have very interesting properties. The first is that, once the aqueous solution on the inside is separated from that on the outside, there is no guarantee that the two will remain the same, any more than the helium inside a balloon has to be the same as the air outside. For instance, you could change the salt concentration of the outside fluid. Typically, salts cannot move without tricks through lipid (fat) membranes, while water frequently can. If you add distilled water to the outside, the water will move into the coacervate bubble, diluting the salt inside of it and causing the bubble to swell. If you add extra salt to the outside, the inside water will move outside, trying to dilute the external salt solution. This is osmosis. Note however that the bubble will change size.

Changing size has consequences for the coacervate bubble. The fats have particular shapes and alignments in the bubble, symbolized by the T shapes in the bubble in the diagram. There is a comfortable size for the bubble, as indicated in the left drawing. If it loses too much water, the lipid molecules finally crowd each other, creating strains as in the center drawing. If it takes in too much water, the lipid molecules spread further apart, weakening the structure until it eventually bursts, as in the right drawing. The result is that, other things being equal, coacervates tend to drift to an ideal size: too small, and they accumulate water; too large, and they split into two smaller bubbles, which will then tend to accumulate water and expand. In other words, they grow, divide, and reproduce. They do not truly reproduce, because we have said nothing about them making copies of themselves that are identical in any fashion except size. A living creature has to be able to control what is inside. However, coacervates can have a stable size that, intriguingly, for fats similar to those found in our cells, is approximately the size of bacteria. And the separation of the phases (the liquid inside the bubble and the liquid outside of the bubble, together with the lipid phase) offers the possibility of controlling what is inside. It is extremely important to control what is inside the bubble, for the proteins with which life is built are very unstable. Proteins cannot tolerate major excursions in temperature, and their shapes change, rendering them inefficient or ineffective, with only modest changes in acidity, salts, or even concentration of salt. To stabilize the inside of a bubble requires catalysts, or agents that speed up reactions, because rebuilding complex agents spontaneously is going to be too slow and random, and, if we are going to replicate this bubble, we will have to copy everything that is inside the bubble to make a new one. The catalysts inside our bodies are almost exclusively enzymes made of proteins. Proteins are long complex chains of amino acids, and they act as enzymes by being folded into complex shapes. The complex shapes manage to fit two or more other molecules, bringing these latter so close that they can interact by exchanging bonds. This is the function of the catalyst: two molecules that would by randomly bouncing around encounter each other only rarely are forced together, causing the reaction to occur much more rapidly. Some enzymes act like cocked springs, temporarily transferring their energy to the reacting molecules so that the reaction can take place, and then reacquiring their energy from another source. Imagining the conversion of a coacervate bubble into one filled with enzymes that can control what is inside and what goes on inside runs into a major problem: proteins are so complex that the spontaneous appearance of just the right form is unimaginable. Leslie Orgel[i], Francis Crick, and Carl Woese had suggested, beginning in the 1960's, that RNA and DNA were structurally much simpler than proteins. If nucleic acids could carry out catalytic functions, then one might conceive of an early form of life in which nucleic acid reproduced nucleic acid, under conditions existing within the bubble, without using protein catalysts. For instance, the bubble might regulate acidity by being differentially permeable to acids and bases. Under this scenario, proteins would be a later development. In the 1980's Thomas Cech and Sidney Altman demonstrated that indeed, RNA, a macromolecule of relatively simple structure, could catalyze reactions. Thus it became at least theoretically possible to imagine a universe in which the first organisms consisted primarily of RNA reproducing itself. Among these organisms, those that began to use DNA, a modified version of RNA that is chemically more stable, would be selected for, leading to organisms in which RNA was the primary catalyst and DNA the repository of the primary information. Finally, among these creatures those in which the RNA could occasionally produce proteins instead of RNA would gain an advantage, because proteins offer so many more possibilities, being able to be folded into an infinite number of shapes to carry out an infinite variety of reactions. Some of the proteins could even dissolve into the surrounding membranes and regulate what comes into and leaves the cell. This of course is all highly speculative, and we have no direct evidence that life appeared in this manner. Such events would have occurred 3.5 billion years ago in a very different atmosphere in which no bacteria were around to destroy whatever was formed. Very little of the ancient earth is available, untransformed, for us to study. However, it is completely consistent with what we can learn about early life. There is chemical evidence for the existence of life before any complex creatures are seen; looking at ancient rocks with electron microscopes reveals patterns consistent with bacterial growth; and the chemicals that can be identified are consistent with the biology of bacteria. Some of the most primitive bacteria that we know—more about that later—can live in conditions similar to those that presumably existed in the ancient earth. These primitive bacteria include the *Archaea* which, though similar in size to bacteria, are now considered to be completely different. Far more numerous than was ever imagined, they were first identified in very hostile environments such as hot springs and high concentrations of salt, conditions that may resemble the conditions of the early earth. They are distinguished from bacteria by their biochemical processes, some of which are unique and some of which resemble those of more complex intact cells (eukaryotes), such that some hypotheses put them at the origin of life on earth and others suggest that archaea, engulfed by other cells, became the nucleus of eukaryotic cells. They include the organisms that produce methane gas. Although it is impossible to identify them by structure in the earliest rocks, the unique chemistry of their cell membranes can be detected, indicating that this group of organisms has existed on earth for 3.5 billion years.

Looking for the earliest signs of life

There are many ways in which we can identify the earliest signs of life. Some fossil formations, called stromatoliths[ii], are remarkably similar to formations made by modern marine bacteria. There are even peculiar tracks in fossils that are similar to those of a recently discovered giant protozoan[iii], which sort of rolls across the deep ocean floor. Pure carbon is not commonly produced chemically. We usually find carbon oxidized (carbon monoxide and carbon dioxide) and further reacted with positively charged ions (calcium carbonate or chalk and other minerals). Living organisms however can decay into carbon-hydrogen complexes (oils and gases) and to nearly pure carbon (coal); the oil coal fields indicate previous life. In even older rocks, there are more subtle clues, based on the fact that the chemistry of living forms is considerably more picky than ordinary inorganic chemistry. One isotope of carbon, ^{13}C, is marginally larger than the more common form, ^{12}C. This difference is unimportant in inorganic chemical reactions but, because reactions carried out by living organisms depend so much more on the shapes of the molecules, living organisms discriminate slightly against the larger form. Thus, in carbon-containing rocks, a slightly lower concentration of ^{13}C than would be expected indicates that the deposit was formed from living creatures. The importance of shape has a further consequence. Many bonds between carbon and other atoms, once formed, cannot move. Thus you can have a configuration like lactic acid, which contains three carbon atoms in a row. The four bonds or attachments of the middle carbon are each different: one hydrogen, one hydroxyl (a hydrogen and oxygen combination), a carboxyl or acid group, and a methyl group. (The latter two contain the other carbons.) In inorganic reactions, this does not make much difference but, because enzyme-dependent reactions depend on the shapes of the molecules, whether the molecule is the left-handed glove or the right-handed glove is very important. To force the reaction the other way would be about as comfortable as putting tight-fitting shoes on the wrong feet. So, perhaps because of the way in which we evolved, by and large living organisms vastly prefer one form to the other. In particular, in proteins, the building block amino acids come in two forms, called L and D (for the Latin words for left and right, because they bend light in different directions). If one causes amino acids to undergo inorganic chemical reactions, it does not make much difference whether one uses the L form, the D form, or a mixture. However, if an enzyme uses an amino acid, except in a few instances it will use only the L form; the D form is the shoe for the wrong foot. To return to the subject, if for instance in a meteorite or very ancient fossil rock we find a 50:50 mixture of L and D amino acids, the likelihood is that the amino acids were formed by chemical means. If, however, we find a predominance of L amino acids, the likelihood is that they came from a living creature. Taking all of this evidence into account, we have a very strong argument that life originated on Earth approximately 3.5 billion years ago. This is so close to the time that the Earth became cool enough to support life that we can speculate that, given the right conditions, life can generate rather easily. This provides the impetus for our search among the heavens for planets that could provide the right conditions or, in the words of the astronomers, "in the Goldilocks zone"—neither too close nor too far from their sun, neither too big nor too small.

This, however, produces a different problem. If life arose soon after the earth became habitable, it was quiescent for an extremely long time. As the graph below illustrates, assuming that the earth is approximately 4.5 billion years old and has been habitable for 3.5-4 billion years, and that the first signs of life appeared 3.5 billion years ago, then why did it not really take off until approximately 500 million years ago?

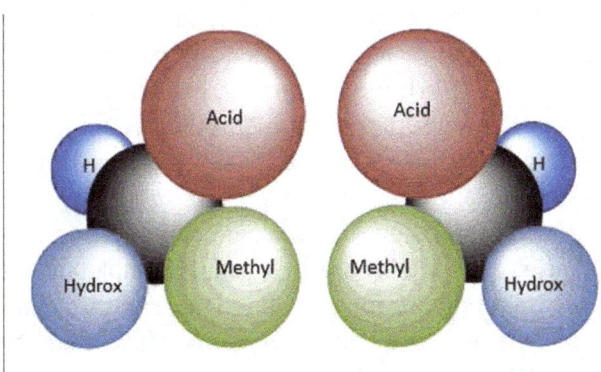

Figure 8.4 Enantiomers. The two molecules are mirror images of each other, like a left and right hand. Biological systems vastly favor one form over the other, though purely chemical reactions do not.

Multicellular organisms

Bacteria can do a lot, but they have their limitations. All of their functions take place within the confines of a single membrane, so that all enzymes must function best at approximately the same level of acidity and in the presence of the same salts. This can be cumbersome. For instance, it is not very practical to store one's food, eat it, and destroy the garbage all at the same temperature. Also, everything—foods, waste products, oxygen, carbon dioxide—must diffuse through the cell membrane, limiting the size of the cell. Life could be more efficient if, first, cells could subdivide their tasks, for instance synthesis, storage, consumption of foodstuffs; synthesis and degradation of cell proteins; and synthesis of DNA; and second, if individual cells could be assigned specific tasks, such as movement, storage of foodstuffs, and digestion. Higher organisms do this: The DNA replication is confined to the nucleus, photosynthesis in plants is in the chloroplasts; respiration (processing of oxygen) in both plants and animals occurs in mitochondria; proteins are synthesized on ribosomes and in the endoplasmic reticulum; and much of the cell's own digestion takes place in lysosomes. Most of these structures are encased in their own membranes and maintain their own environments. For instance, our stomachs produce a lot of acid because it is easier to break down food under acid conditions. Similarly, lysosomes inside cells are typically much more acid than the rest of the cell. Lynn Margulis was the first to suggest how we got these organelles and moved from bacteria, which have no intracellular organelles, to eukaryotic organisms which, as the Greek name indicates, contain a nucleus and other organelles.

It was known that all cells use ribosomes to make proteins but that the ribosomes of bacteria were slightly different from those of eukaryotic organisms. This difference allows us to use some antibiotics that are much more toxic to bacterial ribosomes than to cell ribosomes. It isn't a perfect arrangement, however, because the ribosomes inside mitochondria and chloroplasts are similar to those in bacteria. Finally, by the 1960's, DNA was found inside mitochondria and chloroplasts, and these organelles were seen to divide as the cells divided. Mitochondria and chloroplasts have a third peculiarity: unlike other organelles, they are surrounded by not one but two membranes, a deeply folded inner membrane and a separate, smoother outer membrane. Margulis put it all together by suggesting that these organelles in eukaryotic cells were the vestiges of bacteria that had been captured and were now resident in the cells[iv]. The argument is that certain bacteria evolved to conduct photosynthesis and others (obviously later) to use oxygen to break down sugars completely to carbon dioxide and water (respiration) as opposed to rearranging the sugars to lactic acid or ethanol. These bacteria were obviously very successful. Other cells apparently profited by being near them. Some obviously ate these bacteria, living off their production. Others apparently took them into vacuoles, as if they were going to eat them, but instead stored them within the cells, allowing their useful products to diffuse into the host cells. This arrangement was apparently beneficial to the photosynthetic and respiring bacteria as well—one can imagine several reasons—and it evolved into a mutualistic arrangement in which both partners benefited from the arrangement. It has since evolved into a state in which neither partner can exist alone. The outer membranes of mitochondria and chloroplasts are thought to be the vestiges of the enclosing vacuolar membrane, while the inner membranes represent the old cell membrane of the bacteria.

Since Margulis first broached the hypothesis, considerable evidence has been produced to support it. Briefly, the main evidence is as follows:

- The cytochromes, necessary for respiration, and chlorophylls, necessary for photosynthesis, are embedded in the cell membranes of their respective bacteria, and are embedded in the inner membrane of the organelles within cells. (Actually, this is a practical idea. The flying electrons associated with the capture of energy can do a lot of damage to cells, and are best controlled if all the molecules that do the work are rigidly in place, like a fast-moving basketball game.)
- Mitochondria and chloroplasts each have their own DNA, like bacteria and, like bacterial DNA, the DNA is circular, with no open ends. The DNA carries some but not all of the genes to make the mitochondria or chloroplasts; other genes are conventionally in the nuclei of the cells, so that the organelles cannot survive on their own.
- Mitochondria and chloroplasts can divide like bacteria.
- The ribosomes of mitochondria and chloroplasts are more similar to those of bacteria than they are to the other ribosomes in eukaryotic cells.
- Finally, the base and amino acid sequences of the organelle RNA and proteins have been determined. Not only are they similar to those of bacteria, the families of the bacteria to which they are most similar have been determined. These are photosynthetic and respiring bacteria.

The second important achievement was in making multicellular organisms. There are several advantages in being multicellular. Larger size means that it is easier to store resources. The larger an organism is, the less surface it has per unit volume. It loses or gains materials across its surface but uses them according to its volume. Thus, up to the size where diffusion limits what it can deliver to its interior, being larger means less work in trying to maintain the inside against the outside. For organisms that must depend on diffusion, this limit is of the order of millimeters. But the other important advantage is that it will be possible to specialize individual cells to do specific tasks. The situation can be best explained by looking at an animal that has tried to make a compromise. In coelenterates (sea anemones and hydras) there is a cell called a musculoepithelial cell. It is a skin cell with all the properties of a skin cell, including cilia or little beating hairs.

However, it is also a muscle cell that helps the animal contract when frightened. The problem is that the contracting muscle part must tow along the non-contracting epithelial part, resulting in strain on the epithelial part and limitation of the movement of the muscle. It would be much more efficient if the muscle and the skin were two different entities.

There are some creatures today that give us a sense of what the first multicellular creatures might have looked like. Creatures such as sponges are aggregates of cells, very similar to free- swimming single cells, that stick together to build a larger structure in which the movement of the cilia of individual cells creates a flow that brings single-celled creatures to them, on which they feed.

New sources of energy: photosynthesis, oxygen, and respiration

There are several thoughts on the matter, but most evolve around the issue of energy. Part of the definition of life is that living organisms can take energy to create order—that is, to build the complex structures that they require. In nature, order tends to deteriorate. A watch or computer may eventually fail, but we cannot create a watch or computer or even repair one without providing energy to do so. Our manufacturing typically runs on electricity, which is generated by burning fossil fuels, using falling water to turn wheels, or controlling thermonuclear reactions. In the ancient oceans, without other organisms to degrade spontaneously formed organic molecules or oxygen to destroy them, it would have been possible for the first organisms to extract energy by processing these chemicals. Yeast today can get energy from sugar by simply rearranging it. However, even though the world is vast, this is not a permanent solution. Ultimately, living organisms will expand enough to consume chemicals faster than they are generated, and the whole system will grind to a halt. It will be better to find a more permanent source of energy. It is also not very efficient. Yeast can rearrange sugar to alcohol and carbon dioxide, but alcohol readily burns, indicating that there was considerably more energy to be had.

One source of energy would be sunlight, which will continue for the indefinite future. There are however two problems with sunlight. The first is that it is very powerful. As we know, it can bleach clothes, cause sunburns, cause cancer, and kill bacteria and molds. This is because the energy contained in each photon is enough to shatter molecules. The second is that we cannot put light into a bottle. In other words, we cannot store light energy. To use light energy, we have to do the equivalent of catching a bullet in flight, without letting it harm us, and deflect the energy of the bullet so that it does work, such as turning a motor. These two tasks were accomplished by the first organisms that achieved photosynthesis.

Photosynthesis is a process by which solar energy (the photo- part) is used to combine carbon dioxide and water into sugar (the –synthesis part). The genius of this is that, while light cannot be stored, sugar can be, and it can be transported elsewhere in the organism to be used when and where it is needed. When sugar is consumed and reconverted to carbon dioxide and water, it releases that energy, which can be used to perform other tasks. The cellulose of wood or paper is essentially a polymer of sugar, and the energy that is released when it is burned is the same as that which, more slowly and in a more controlled fashion, is released when we consume sugar in our bodies. The heat of our bodies is produced by the burning of sugar.

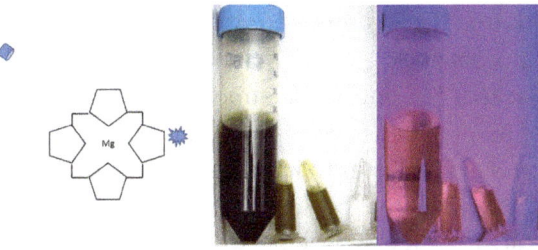

Figure 8.5. Left: In photosynthesis, a high-energy photon slams into chlorophyll, which, being somewhat flexible, absorbs the shock before releasing the energy shortly thereafter. Through several steps, this released energy is used to synthesize ATP. See animation[v]. Right: Chlorophyll, extracted from spinach into hexane on the left, produces the predominant green color. When this solution is viewed under blacklight (long wavelength ultraviolet) it glows red. What has happened is that the chlorophyll has absorbed the ultraviolet light and, with some inefficiency and loss of energy, re-emits it as longer wavelength, lower energy light. In the plant, where chlorophyll is coupled to proteins, instead of re-emitting the light it hands the energy to other molecules, synthesizing ATP ("biological gasoline") in the process.

To capture light energy we need what is called a resonant molecule. Because of peculiarities in its structure, it has a certain flexibility and in particular can exist in different stable states, rather like an oil drum lid that can snap between an up and a down position. They can absorb light at certain wavelengths and therefore are colored ("chlorophyll" means "the green in leaves", their color being the wavelengths that are not absorbed but rather are reflected. When a photon of the appropriate wavelength slams into them, rather than shattering they recoil like an elastic curtain or a trampoline, holding the energy (now measured as the displacement of an electron) briefly before releasing it, as a trampoline springs back. See animation. This exchange is not 100% efficient and, like a hot potato game or a person on a trampoline, some energy is lost. The result is that the electron now bounces to the next resonant molecule with a little less energy, and continues until it reaches a minimum at which the energy can be transferred to attach a phosphate to a molecule called ADP to create an energy-rich molecule, ATP, a sort of biological gasoline, that can be taken elsewhere and used to make sugars or other products. Some more is used to attach hydrogens to a carrier, termed NADPH. The hydrogens will be required to complete the synthesis of the sugar. This is the -synthesis part of photosynthesis. You can actually see this effect. If you take leaves rich in chlorophyll, like spinach, and extract the chlorophyll into acetone (fingernail polish remover) or hexane, you get a clear green solution. If you shine a blacklight (near ultraviolet) at it in a darkened room, it will glow a deep red color. This Halloween trick demonstrates that the chlorophyll captures the ultraviolet light and re-releases it as lower-energy red light. In the leaf, the released energy would instead be captured and turned into chemical energy.

In the early seas there was a lot of hydrogen sulfide, H_2S. The energy used from capturing the photon can be used to strip the hydrogens off the hydrogen sulfide and attach them to carbon dioxide, creating sugar: $6\ CO_2 + 12\ H_2S \rightarrow C_6H_{12}O_6 + 12\ S + 6\ H_2O$. Other than the sugar, there are two waste products: water, which is innocuous, and sulfur, which is insoluble and will sink to the bottom of the ocean or lake. This is not very harmful at first, but eventually it will accumulate. The sulfur deposits underneath salt domes are thought to have accumulated in this fashion. This is one problem: the waste material will eventually choke off the lake. A second problem is that H_2S enters the atmosphere from volcanoes and is a limited resource. It (happily) exists in relatively low concentration in the air and, if life expands, can be used up faster than it is produced. Using photons to split hydrogen sulfide and was a great advance—it produced enough bacteria to fill lake beds with sulfur, and there are plenty of bacteria today that can carry out these reactions—but ultimately it was limiting.

The next great advance was to find a similar but very abundant material to use in place of hydrogen sulfide. In spite of their substantial differences in properties, chemically oxygen and sulfur are very similar, as are hydrogen sulfide (H_2S) and hydrogen oxide (H_2O or water). This was a brilliant solution. Water was far more abundant, and the waste product, oxygen, would simply escape to the atmosphere. Thus a new adaptation was to carry out the revised reaction: $6 CO_2 + 12 H_2O \rightarrow C_6H_{12}O_6 + 6 O_2 + 6 H_2O$ one molecule.) There was no problem with this solution, other than that it began to pollute the atmosphere with oxygen. We know that free oxygen did not exist in the early atmosphere because it is so reactive, and materials exist in the earliest rocks that would have been oxidized had oxygen been present. Similarly, we can estimate from other chemicals that begin to appear in later rocks when oxygen first appeared, and we can estimate its concentration by the state of these materials.

At first, the amount of oxygen appears to have varied, as we have peculiar banded rocks[vi], in which rust (iron oxide) alternates with pure iron, indicating that, during the formation of the mud that eventually became a rock, oxygen was alternately present and absent. This could have been because of fluctuations in the atmosphere or alternate submersion and exposure of the muds to the atmosphere.

The appearance of oxygen marks the beginning of water-based photosynthesis, coincident with fossils resembling cyanobacteria (blue-green bacteria, the color coming from the photosynthetic pigment chlorophyll), approximately 2.8 billion years ago.

(Oxygen is written "6 O2" rather than "12 O" because two atoms of oxygen travel together as one molecule)

The accumulation of oxygen in the atmosphere had a major impact on the world. Without oxygen in the atmosphere, sunlight is a powerful UV-sterilizing system. The ultraviolet light can kill bacteria today, but in the upper atmosphere it can convert oxygen, two atoms of oxygen combined into one molecule of oxygen gas, into ozone, which has three atoms per molecule, and ozone can absorb and therefore block ultraviolet light. Thanks to the oxygen in the atmosphere, between 98 and 99.2% of the ultraviolet light hitting the upper atmosphere never reaches the surface of the earth. (The number varies because cloudiness, humidity, sun angle, and elevation all affect the measurements.) Suffice it to say that without oxygen the surface of the earth would be one big germicidal bath, and life could exist only under water, since water also blocks UV. Again relying on the types of compounds generated when oxygen exists at specific concentrations, as well as, in later evolution, direct analysis of gases trapped in rocks and minerals oxygen (which today is 21% of the atmosphere) appeared in the atmosphere 3.6 billion years ago, and gradually increased. By one billion years ago, it was up to 15% of the atmosphere, and reached near 21% by the beginning of the Paleozoic period, about 500 million years ago (Fig. 8.6). The other important aspect about oxygen is that it reacts so well with everything. Prior to the appearance of oxygen, the only way to get energy to build new organisms was through fermentation: the rearrangement of sugar molecules to form lactic acid or ethanol, carbon dioxide, and water. This works, but both ethanol and lactic acid burn, indicating that there is a lot more energy available. Burning of course is reaction with oxygen. Once oxygen is available, it becomes possible to burn—slowly and under full control—sugar completely to carbon dioxide and water, yielding 18 to 24 times more energy. So oxygen in the atmosphere both allows organisms to climb onto land and makes it possible to gain far more energy from the consumption of resources. An increase in energy yield means smaller storage required and greater activity possible, liberating creatures to acquire new shapes and become more agile.[vii]

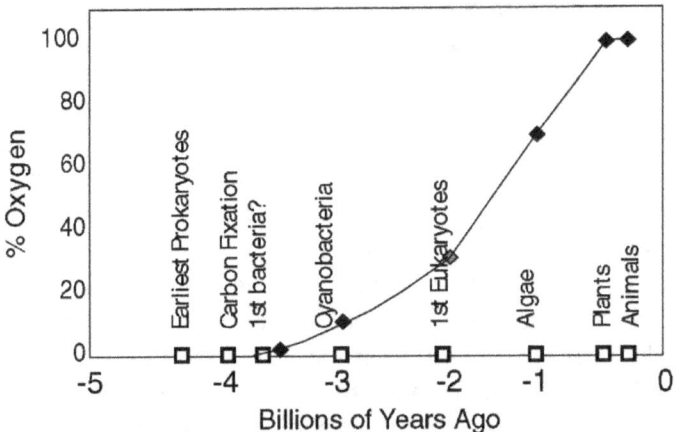

Figure 8.6. The appearance of oxygen on the planet

Chapter 9: Ages of Earth

WHO KILLED THE DINOSAURS? THE CHICXULUB DETECTIVE STORY, OR CSI INVESTIGATES

The end of the dinosaur era seems to have been rather rapid. In evolutionary terms, this can mean hundreds of thousands of years, but even in these terms there is evidence of massive die-offs, for instance in the bends of former rivers in the American Great Plains, where huge numbers of dinosaur bones washed ashore and accumulated. There had been much speculation about various causes, ranging from the rise of mammals—not likely, since mammals were not-very-impressive small, rodent-like creatures at the time, and in any case the rise of numerous variations on the mammal theme is a typical adaptive radiation or expansion of mammal varieties to fill the niches left by the already-gone dinosaurs. Other explanations included temperature changes: if the earth got too cold, a large, cold- blooded reptile might never get warm enough to move easily; or availability of water might have been an issue. Whatever happened, there was a massive change. Something like 57% of all land species died very abruptly. The change was so sudden that there is a shift in the appearance of the soil, owing to a drastic reduction in carbon content (from decaying animal and plant life) at this period, and the transition is known as the K-T boundary (from K for Cretaceous or chalk-bearing, and the K used to distinguish this era from the earlier Carboniferous or carbon- bearing era; and T for Tertiary, or the beginning of the era of mammals) Thus the world had changed abruptly from one filled with reptiles and tree ferns to a more impoverished one and finally to one dominated by mammals. The detective story begins with the effort of Walter Alvarez to get his father, Luis to help him understand a question of geology. As is described in Chapter 4, one can date rocks using radioisotopes and composition, but there are occasional anomalies. For instance, the metal iridium, which can be used to assess certain dates, is relatively uniformly and rarely found in the earth's crust, though there is much more deep within the earth. In 1973 Walter Alvarez enlisted his father to help understand why, in a region of Italy, there was an exceptionally high deposit of iridium, hundreds of times higher than normal. What they determined through exploration was three observations, which they summarized in 1980:

• The iridium anomaly was found in a very thin layer of the earth, rather than being generally distributed;

• The same anomaly could be identified not only in Italy but far from Italy, in Spain and even in the Americas;

• And, the iridium layer could be dated to 65 million years ago—the same date as the K-T boundary, or the disappearance of the dinosaurs.

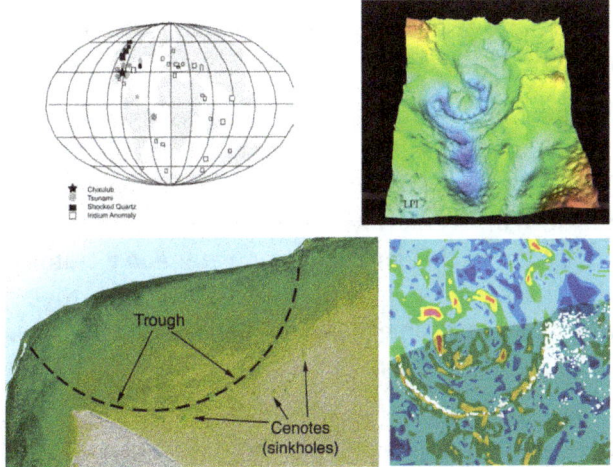

Figure 9.1. Chixulub. (Upper left). Map of locations of evidence of a tsunami (along the American coasts); tektite fallout, which are in a gradient with larger tektites being found to the north and west of the Yucatan peninsula; and iridium anomalies, which are world-wide but more equatorial than elsewhere. The map illustrates the positions of the continents at the time of the Cretaceous mass extinction, with India not yet having collided with Asia. The approximate site of Chixulub is marked with an X[i]. (Lower left Evidence of the crater at Chixulub. There are traces of a semicircle on the north coast of the Yucatan peninsula, with sinkholes all along the semicircle. The sinkholes were caused by shattering of the coral structure with the impact[ii]. (Upper right): magnetic anomalies, also consistent with the remnants of a meteorite[iii]. (Lower right): A computerized image of the variations in gravitational fields[iv], a measurement of the density of the soil and rock and therefore a measurement of the total metal and compression of the soil, along the coast. Today's coastline runs from left to right approximately through the center of the impact zone. All the evidence is consistent with a collision of a meteorite coming in from the southeast and plowing toward the northwest.

Figure 9.2. *Left.* The K-T (Cretaceous-Tertiary) boundary is rather sharp and is marked by an increase in carbon content in the soil directly above a low-carbon layer filled with increased iridium and shocked quartz, above which is sharp decrease in the carbon content of the soil, reflecting a decrease in the number of organisms alive.[v] Right: the sudden decrease in carbon (black coal-like soil) is often very apparent.[vi].

There is another point that they understood, which applies to this situation. Iridium is quite rare on the surface of the earth, but much more common in meteorites. They therefore made the following hypothesis: A giant meteorite had struck the earth 65 million years ago, scattering iridium as dust throughout the world. This meteorite had thrown up such a dust cloud that it seriously interfered with photosynthesis, disrupting the ecosystem of the planet and causing massive die-off of dinosaurs and many other organisms. The minerals in the dust would have been converted into acid, creating acid rains that would have further seriously damaged the ability of organisms to survive.

This was a very interesting hypothesis, but there was no other evidence for it. The question therefore became whether one could find independent evidence for such a meteor strike. The problem was that there were no craters on earth that could be linked to such an event. However, this was not a fatal argument or true falsification, since 2/3rds of the earth is under water and a crater might not be noticed; or it might have existed in now-eroded lands. Nevertheless a hypothesis is interesting when it suggests an experiment and the question was then what other evidence of a meteorite might one expect to find.

If a giant meteorite were to hit the earth, the shock would be immense. In 1908 a meteorite, estimated to be approximately a football field in diameter, entered the atmosphere and exploded at a considerable height above the earth over Siberia. The sound of the Siberian meteorite exploding could be heard for hundreds of miles. It flattened trees over 800 square miles. Likewise, the explosion of Krakatoa produced shock waves and tsunamis that were registered on machines around the earth, and the sound of the explosion was heard 2000 miles away. Such powerful events shatter and melt rock, and therein lay our clues, in the form of tektites and shocked quartz. The Alvarez hypothesis was so compelling that a search began for such further supporting evidence of a meteorite.

The first item sought was tektites (Fig. 9.2). If you were to take ball of molten glass and to drop it into a tub of water or otherwise cool it very quickly, the outer glass would cool very quickly into a hard, rigid, shell, while the inside would cool more slowly and either have to conform to the shell or fit uncomfortably within it.

Such a phenomenon is exploited in the construction of the glass insulators that hold the high voltage electrical lines to telephone poles. In this case the interior glass exerts considerable pressure on the shell. If the shell cracks, the whole insulator shatters. This makes it much easier to spot from the ground a defective insulator than if, for instance, the insulator were ordinary glass that could fail by developing a small, inconspicuous crack. In any event, when examined by appropriate microscopy, the shell should be distinguishable from the inner, more slowly cooling, mass. If a meteorite hits a rocky surface, it will be hot enough and exert enough force to melt the rock, such that sand could be converted into glass and other minerals into similar glasslike substances. These will be ejected, in molten form, from the impact site and will cool rapidly in the atmosphere, forming such a shell. Small stones formed in this fashion are called tektites, and they mark sites of ejection of molten materials, such as volcanoes and impact sites. Even more interesting, depending on how high they go into the atmosphere and how rapidly they cool, they may take on teardrop shapes. If the place in which they land is completely undisturbed, they may even lie in an orientation that suggests their origin. If the trajectory was relatively low and long, their pointed tails will face the direction from which they come. Also, the smaller, lighter ones will fly farther than the larger, heavier ones, and one can trace the source if there is a geographic distribution in size. Tektites are common and had not been much studied, but a second look demonstrated that many, particularly in the Caribbean, seemed to be about the right age of having been formed 65 million years ago. That turned much more attention to the Caribbean.

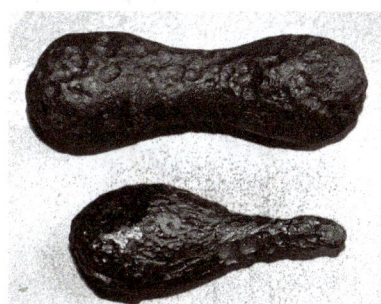

Fig. 9.3 Tektites[vii] showing the characteristic tear shapes. Some are much rounder.

That turned more attention to some peculiar deposits of sand well inland in Texas. good evidence for a lake or an ocean in that region. The distribution of sand however suggested a violent origin such as a tsunami or other sudden inrushing of water. But tsunamis come a few or tens of miles inland, not hundreds of miles inland. There was another peculiarity with the Texas sand. It was shocked quartz. You are familiar with the situation in which a piece of glass, particularly a hard glass such as Pyrex®, breaks. A small crack forms, and this crack propagates rapidly, branching and spreading until there are cracks all over the glass. Quartz is very similar to glass, and even harder. The atoms in quartz tend to be aligned, so that polarized (aligned) light shining through quartz looks different according to the angle. When the alignment of the quartz is in the same direction as that of the light, light passes through. When the alignment is at right angles, the light is blocked, and at intermediate positions, one sees different colors. (You can see the light-blocking effect by taking two polarized sunglasses and rotating one relative to the other while looking at the sky. The polarizers will become opaque when they are "crossed" or at right angles to each other.) The value of this insight is that polarized light allows us to see the structure of quartz. Using this technique, we can see if the quartz has been shattered, or is "shocked quartz". It still holds together, but the crack lines are there (Fig. 9.3). The sand in Texas is not only in the wrong place, it is shocked. What this suggests is that a powerful impact produced both the shock and a huge tsunami that flooded Texas. Again, the various dating mechanisms suggested a date of 65 million years ago. In the eastern Caribbean, tektites suggested a source to the west, and the flow lines of the presumed Texas tsunami pointed eastward, drawing attention to the Yucatan Peninsula in Mexico. But there was no crater in the Yucatan. Or was there? From the surface, none is apparent, though there is a modest quarter-circle of a small ridge on the north coast. However, the Yucatan is basically an old coral reef, easily eroded, and sands can easily shift and fill holes in the sea floor. More sophisticated techniques were obviously necessary. These included forms of radar and sonar, techniques that send microwave and sound wave signals, respectively, and listen for the echoes. Harder material will send a more identifiable echo, and the timing of the echo will indicate its distance or depth. These techniques produced a very surprising result: Although it was filled with sand, there was a very clear circular outline along the north coast of the Yucatan Peninsula.

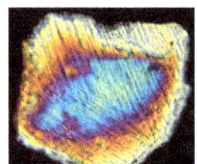

Figure 9.4. Shocked quartz [viii], as seen through crossed polarized light. The fractures in two directions are clearly visible. Material such as this is found deep into Texas.

The crater is approximately 100 miles across, suggesting the impact of a meteor approximately 4 ½ miles across. It is approximately 65 million years old, and it has another interesting property: judging from the structure of the crater, the meteorite came in at a very shallow angle from the south-southeast (Fig. 9.1). The size and angle lead to several interesting predictions. First, it could have generated a Texas-bound tsunami of sufficient size to carry the sand appropriately far into Texas. Second, the heat generated would have been sufficient to generate a firestorm over much of North America. From various lines of evidence the dying of the dinosaurs appeared to begin in North America and then to spread to Europe and Asia. The direction of the meteorite could justify this argument. Thus this beautiful detective story meets our criteria in terms of **Evidence**, particularly multiple, independent, sources of evidence, and **Logic**, in that there is a reason for the dinosaurs to die. If the explosion of Tamboura or Krakatoa was bad, this would have been much, much worse. Sufficient dust would have been ejected to seriously undermine photosynthesis for at least a decade, leading to massive starvation and collapse of ecological cycles of dependency. The loss of larger numbers of land than marine animals would likewise follow, since the ocean would have at least been protected from fires. To summarize the evidence and logic:

- There is substantial evidence that a giant meteorite hit the earth 65 million years ago. The iridium layer, the tektites, the shocked quartz, the evidence of a tsunami, and a crater all point to an impact site on the coast of the Yucatan Peninsula in Mexico

- There is substantial evidence for a massive die-off of reptiles and many other organisms 65 million years ago, so substantial that it marked the end of one era and the beginning of another.

- Logic—that is, calculations—indicate that a meteorite big enough to make the Chixulub crater could have created a firestorm and a dust cloud large enough to strongly decrease photosynthesis and could cool the earth enough to make life very difficult for cold-blooded creatures. Certainly the changes to the climate would have done great damage.

ARGUMENTS AGAINST THE CHICXULUB HYPOTHESIS

What we do not have is true falsifiability. Obviously, we have no desire to replicate the experiment, but is there any way that we can test the hypothesis? Not really, but there are those who do not consider what is now called the Chicxulub hypothesis to be correct. They base their arguments on four issues: The calculations all are based on many assumptions, and there is some disagreement over the assumptions. For instance, dust in the atmosphere will cool the earth, but so-called greenhouse gases will warm it. (Greenhouse gases act like one-way mirrors: they allow sunlight to reach the earth but, when the sunlight warms objects on the surface of the earth and the heat rises into the atmosphere, the gas reflects the heat back to the earth, trapping the heat. To be more technical, they are transparent to light, allowing it to reach the earth from the sun, but opaque to heat, reflecting it back toward the earth.) How much the two effects will cancel each other out is a matter of dispute. An impact the size of this meteorite would certainly have repercussions around the world. The immense lava flows of Siberia and India were generated at approximately this time. Perhaps the meteorite triggered this activity, perhaps not; but in any event the volcanic activity could have generated the dusts and the greenhouse gases without invoking a meteorite.

There was evidence for a decline in dinosaur populations before the K-T transition. It is not clear what caused this decline, or whether the populations were already in severe trouble. Thus the impact might have been the death knell for a deteriorating pattern of life, or have had minimal impact on an already imminent collapse. We cannot tell if the "rapid" collapse occurred in weeks or over thousands of years.

The K-T event, while spectacular since it is relatively recent and involved gigantic animals, is only one of several population collapses, and it is far from the most massive. There is no solid evidence that meteorites triggered the other collapses. The detective story of "Who killed the dinosaurs" illustrates very well the limitations of science. We have a wonderful story, complete with evidence and logic, that a meteorite wiped out the dinosaurs. Today it is enshrined in children's tales and movies. Just because it is appealing however does not make it so, and many questions remain and must be answered before we can really consider this hypothesis to be a theory. The most important consideration is that of the other massive declines in life on earth (Fig. 9.5). Although people are looking for evidence of meteorite impact at these times, the further back in time one goes, because of erosion, continental drift, and other processes that replace or alter exposed rock the harder it becomes to document so transient an event as the impact of a meteorite. We are left to conclude that the impact of the Chicxulub meteorite was too close to the crash of the dinosaurs for it to be entirely coincidental, but we cannot prove it. We are of course very interested in what causes these crashes, from both the intellectual and purely selfish viewpoints. The loss of 80% of the weirdest possible creatures cleared the way for the rise in recognizable life forms that we know as the Cambrian, the beginning of the Paleozoic; these creatures thrived for almost 300 million years before the amphibians died off (the Permian mass extinction, eliminating up to 60% of existing families) and left an opening for the reptiles, marking the end of the Paleozoic and the beginning of the Mesozoic (some people claim to have located a crater representing a meteorite "the size of Mount Everest," but this finding is disputed); the reptiles dominated the earth for approximately 120 million years before the K-T extinction swept away at least 20% of all existing families of organisms; and in their swath, the mammals rose to characterize the Cenozoic. If we look to the future, we have to ask: Will we die off? What might cause us to collapse? What might replace us?

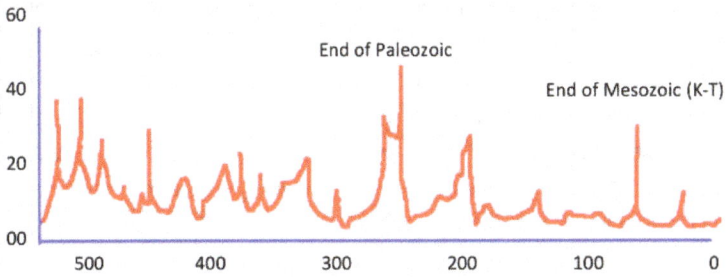

Figure 9.5. Mass extinctions. 75% of all existing families died out at the end of the Precambrian and the beginning of the Cambrian (at left, before 500 million years ago (mya)). The disappearance of over 60% of all families marked the end of the Paleozoic in an event termed the Permian mass extinction (label in middle, approximately 250 mya). In contrast, the Cretaceous mass extinction, ending the reign of the dinosaurs and marking the transition from the Mesozoic to the Cenozoic, though spectacular from our viewpoint, was far less drastic than the earlier ones, perhaps because the total variety of creatures on earth has been steadily increasing. Ordinate: Percent of marine species that went extinct. (The percentage on land would likely have been higher.) Abscissa: mya. Redrawn from http://upload.wikimedia.org/wikipedia/commons/0/06/Extinction_intensity.svg[ix]

The origin of whales

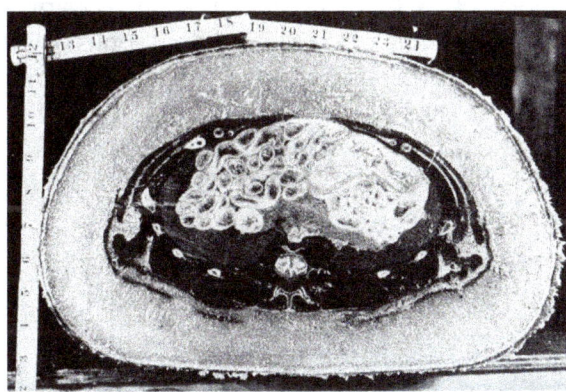

Figure 9.6. Cross-section of a seal[x]. The outer layer (blubber) is all fat insulation.

The 18th and 19th Centuries provided much food for thought for anatomists studying whales. After considerable debate as to whether they were fish or mammals, finally resolved as mammals, since they had lungs rather than gills and suckled their young, several questions arose: why in their forelimbs (fins) they had wristlike bones and bones that appeared to be five digits, when the limb was a paddle; why their tails moved up and down, rather than side-to-side, like that of a fish; and why, in some whales, one could find a set of bones, apparently useless and supporting no muscle or structure, approximately where the pelvis might be. Gradually a consensus developed that whales were indeed mammals and that they had descended from a type of mammal that had lived on land. The goal then became to find fossil proof of this hypothesis.

Mammals evolved as and essentially are land animals, and they accommodate with some awkwardness to other media. For instance, bats, unlike birds, have teeth. Teeth add extra weight to their jaws, creating a balance problem that must be resolved in a fashion unlike that of birds, who have substituted a stone-containing gizzard (in the abdomen and thus centrally located relative to the wings) in place of teeth, which are replaced by the relatively light-weight beak. Bats also have a very high metabolism and must essentially hibernate so that they do not starve to death during the day. They also must bear their young alive. As in an airplane, the extra weight is very expensive to carry, incapacitating bats in late pregnancy. Birds more efficiently quickly develop and lay a large egg, thus freeing themselves to fly and collect food.

Aquatic mammals have similar problems. Water drains heat far more effectively than air. Fish surrender to the problem, allowing the water to cool their blood in the gills to ambient temperature, but poikilothermic (cold-blooded) animals can be paralyzed by low temperatures and may otherwise be limited in speed. A few large fish, such as tuna, have circulatory arrangements and metabolic modifications (the dark meat of a tuna) to allow them to keep their muscles warmer, giving them more speed. Mammals, however, are built like racing cars and, like racing cars, must remain precisely tuned (warm) at all times. A temperature of 98-99° F (37.5° C), incidentally, is about as warm as one can get—to run metabolic reactions at their fastest—without seriously disrupting protein structure. Aquatic mammals therefore carry enormous layers of blubber to allow them to retain heat in the core of their bodies. Almost 60% of the volume of a seal is blubber.

Fig. 9.7. Water-dwelling animals. Clockwise from top left: tropical reef fish (fish); capybara (mammal); manatee (mammal); orca (mammal); penguin (bird); sea lion (mammal); shark (cartilaginous fish). Cartilaginous fish are not true fish with mobile fins and scales. Note the increasing evolutionary adaptations to life in water their among the various birds and mammals, from capybara to sea lion to manatee to orca, indicated by their increasing fish-like shape and inability to function on land.

Truly aquatic mammals must make other compromises as well. They must breathe air, as there is no way in which they could extract enough oxygen from the water to survive without massively bringing the blood close to the water, as in gills, and therefore cooling it to the ambient temperature. They also must deliver and suckle their young in water. Several seagoing mammals have developed a social structure in which midwives assist at birth. Since the infants have never breathed air, their lungs are collapsed at birth. They therefore would sink in the water, except that the midwife animals lift them to the surface, where they can first fill their lungs with air and therefore become more buoyant. So how did they arise? The story was first well documented from the fossil record. Subsequently, measurements of DNA sequences have corroborated the hypothesis.

Many mammals are comfortable in water. The South American capybara, a nearly pig-sized rodent, a relative of the guinea pig, is fond of water plants, diving at night to forage on them. Other than being able to stay underwater for several minutes, it otherwise lives and breeds on land. Likewise, a beaver, which can easily move on land, has a broad tail adapted to propel itself in water but otherwise lives above water in its nest, which is above water, though its entrance is underwater. An otter is sleek and streamlined, with a flexible body and strong, powerful tail so that it swim rapidly, fishlike, in the water.

Figure 9.8. Young hippopotamus. Although it can run fast, it is not a graceful land mammal.

These animals provide no particular problem concerning the ability of mammals to survive in water. More problematic is an animal such as the hippopotamus (literally, from the Greek, "river horse"), a large herbivore that keeps cool in the hot African sun by spending most of daylight hours in water. It can run quite fast (30 km/h or 19 mph) over short distances but, given its bulk, you would never describe it as graceful. Given that it has to carry up to 3 tons on short, stubby legs, it does not take much imagination to conclude that it could more easily relax in water, where its massive blubber, less dense than water, would give it some buoyancy. Its leg bones are quite dense, helping to support its weight on land and equilibrating it in the water. And therein lies our first clue. One can begin to estimate the lifestyle of an animal by looking at it. We have previously encountered this argument in noting that aquatic and aerial animals, because of the stronger constraints imposed by their environments, tend to have characteristic shapes. Consider a primarily aquatic reptile such as an alligator, crocodile, or caiman. Like otters, they have short legs and powerful tails, which they move from side-to-side in propelling themselves through water.

Finally, we move to creatures that regularly climb onto land but are much more adapted to water: sea lions and their ilk (sea elephants, walruses), which have external ears and rear flippers that clearly are two very modified hind legs; and seals, which have no external ear pinnae and only one rear flipper. (Fig. 9.7) We first encounter mammals that have similar builds in fossils from Pakistan, the land of which has lifted from the sea because of continental drift. They are creatures like Pakicetus (literally, Pakistan whale). Like an alligator, it has dense leg bones, a long snout, and a powerful tail. Judging from its shape, it was a strong swimmer but primarily a land carnivore. It can be related to whales because of some particular aspects of the anatomy of its skull, which are found in no other mammals. This creature, which lived about 50 million years ago, during the rapid expansion of mammals, was apparently a water-going carnivore. We can tell a lot about its lifestyle by studying its skeleton. For instance, its eye sockets were high and facing upward, like those of an alligator. This would allow it to hunt by floating on the water. The ions in its bones, measured by radioactivity, indicated that its habitat was fresh water. About one million years later a more powerful, better swimmer emerged; it was more alligator-like and was given the name *Ambulocetus natans* (the swimming walking whale). Its isotopes indicated that it least occasionally moved in salt water or ate salt-water prey. However, it was a shallow-water creature as indicated by one very important characteristic: its rib cage was complete.

Fig 9.9. Left: Artist's impression of Ambulocetus. Right, a young caiman.

Humans cannot dive unaided to great depths because pressure increases with depth, compressing the air in the lungs and causing them to shrink and collapse. Short of sucking all the organs into the chest, there is nothing to fill the space in the rib cage, and ultimately the ribs will crack inward. To avoid this disaster, deep-diving whales have open rib cages. There is no sternum, and the ribs originate along the spine but do not meet in the front of the chest. Thus the lungs can collapse without breaking the ribs. Whales also have another adaptation. In whales, the inferior vena cava, the large vein that returns blood to the heart from the abdomen, is massive so that, when the whale dives, a huge amount of blood flows into the thorax, occupying the space left by the collapsing lungs. Whales also have many other adaptations, but these are mostly metabolic or exist in soft tissue and so are not easily documented in fossils.

From this point on a series of other fossils was found, more and more like true swimming mammals and extending far from Pakistan, but characteristically in what were shallow seas. It is rather hard to tell when the rib cages of whales opened, because they are often crushed in fossils, but by 35 million years ago there were certainly salt-water, strong-swimming large creatures worldwide. Their nostrils have now moved far enough backward on the head to constitute a true blowhole, while the eyes have now returned to a lateral position. They had flukes, and for some the hind limbs had nearly disappeared, though skeletal remnants of the pelvis and hind limbs remained. They were no longer connected to the spine and, though perhaps they were used in mating, they could not have functioned as limbs.

DNA and protein similarities lead to similar conclusions, that whales evolved from hippo-like animals and are distantly related to pigs. The family started as carnivores, of which some members became herbivores. There is much other evidence as well that whales are related to land mammals. As in many instances of embryology, more mature structures are built on the scaffolding of less mature structures and therefore embryos display the more primitive scaffolding. For instance, fetal whales have body hair. Baleen whales have no teeth but rather, a hard, horny net that filters small swimming animals; their embryos also have teeth. Limb buds and external ears appear in most whale embryos but fail to grow and ultimately disappear. The nostrils start at the front of the head and get pushed back, by differences in growth rate in different axes, to the position of the blowhole

~~~~

Chapter 10: From whence new species?

If we can acknowledge that all available scientific evidence points to the hypothesis that there was an original ancestor of all extant animals and plants, from which we inherited our means of storing and retrieving information to create new individuals very like us, as well as many aspects of the construction of our cells, then we can continue by asking how it is that we got so many species from this ancestor, why there are not other ancestors, and why some species or very similar forms persist much longer than others. We address here some of the mechanisms of speciation. From this point we will look at the history of the planet, and finally at the derivation of our human species from our closest relatives.

Mechanisms of speciation

"Survival of the fittest" refers only to the probability of leaving young to the next generation. As can be demonstrated easily in a laboratory setting and can be surmised by following the fates of certain species, the process is extremely opportunistic and shows no sign of foresight or long-range planning. In evolution, species adapt to the most proximate advantage, sometimes with potentially disastrous results, such as some hermit crabs, which preferred shells of now-extinct snails, adapting to use as shells discarded beer cans littering the bottoms of some bays. Whatever works, works. Strength, speed, or wiliness might be one way to go, but stealth, inconspicuousness, or cowering fear might also allow one individual to survive and thus lead the evolution of the species in that direction. So how might it work?

Selection

Looking at animals rather than plants, one can observe one of the most obvious forces of evolution: predation. Vulnerable animals are commonly consumed by predators, often in horrendously high numbers If you ever watch a fish or other carnivore discover a nest of fish eggs, which can contain up to 10,000 eggs, you will see the attacker excitedly slash its way into the nest, eating everything in sight and sparing only a few eggs it left behind in its slovenly haste. This is the first of the Darwin/Wallace/Malthus argument. Though for eggs you would have to turn to the ability of the mother or father to camouflage them, for the individual hatchling anything about it that would improve its chances of avoiding the hecatomb would increase its potential of survival. This could be anything: slightly earlier hatching; less buoyancy so that it sinks to the bottom; greater oxygen need so that it rises to the surface of the water, away from the other eggs, soon after it hatches; considerably greater skittishness and less curiosity than its siblings; a color or shape that makes it harder to distinguish against the background where the eggs are laid; or something unpalatable about it. It doesn't take much. Anything that improves its survival in this first test and, like most animals, this most vulnerable point in its life, will improve the chance that it will survive to adulthood and breed to pass its characteristics to the next generation. Thus we find creatures with remarkable and even ridiculous adaptations—creatures that one could define, jocularly, as proving the existence of God, because Nature could not have invented something so bizarre. (Fig. 10.1). With just a few requirements, we can mathematically demonstrate that such a selection can change a population as J. S. Haldane did in 1924. If the population is small and randomly interbreeding, and there is sufficient selection pressure, in a few generations it is possible to convert a population displaying one characteristic (for instance, yellow body color in flies) to a population displaying another body color. There must be a measureable selection pressure, meaning that for every one hundred members of the population bearing one color that survive to adulthood, fewer than one hundred bearing the other color do so. It takes little imagination to surmise, for instance, that a juicy caterpillar that happens to be white will be more easily seen by birds than one that is the color of leaves, which is why very few albinos are found in nature, though the loss of an ability to make pigment is a rather easy and common mutation. One can also surmise that the selection would continue from generation to generation, from greenish toward a green just the shade of the leaf toward not only green but patterned and even shaped like a leaf. Some camouflages (or cryptic coloration see Figs. 6.1 and 6.2) are truly startling in their accuracy. But does the camouflage actually work? In 1953 Bernard Kettlewell examined this proposition looking at the peppered moth in England. This moth was known to have a dark mottling on a relatively light background, blending well with the lichens on the barks of trees where it tended to rest during the day. In the 19th C, naturalists noticed the appearance and rapid increase in numbers of a very dark, nearly black variant that, in contrast to the normal light form, was much harder to see on the soot-covered trees that were beginning to dominate the landscape. Raising large numbers of both light and black moths, Kettlewell released them at the edges of woods either covered or not covered by soot, marking them so that he would not confuse them with those already there. He later returned and trapped the surviving moths. Sure enough, far more light moths survived in unpolluted woods, while far more dark moths survived in the polluted landscape. He even did field observations in which he saw birds eating moths that stood out from the bark on which they rested. Subsequent studies have confirmed his findings, in that direct counts of birds eating moths that have been tied to contrasting or matching bark have been undertaken. Furthermore, as industrial soot has been brought under control and lichens are now returning to the bark of the trees, the light form is once again increasing in the population—one of the many instances in which one can watch evolution in action.

Figure 10.1. Some bizarre creatures of this world: Clockwise from upper left: Leaf insect; armadillo; brown pelican; anteater; manta ray; shrimp; pacman frog; praying mantis [i] [ii].

Of course, there is a corollary to this: If you are capable of defending yourself, you want to warn potential troublemakers. Thus toxic or otherwise dangerous animals usually advertise their threat and, even more, they commonly use the same types of warning signals, so that predators have to learn only once. The brighter or showier the defense, the less likely a predator will make a mistake. Both monarch caterpillars and black widows are quite toxic. In fact, Lincoln Brower, using the monarch butterfly (which, like the caterpillar, borrows toxins from the milkweed that the caterpillar ate) demonstrated the effectiveness of this warning or aposematic coloration. Taking a blue jay that had been hand-raised in the laboratory, he offered it a monarch butterfly to eat. Not recognizing its bright orange coloring as a warning, the jay ate the butterfly and, shortly thereafter, retched. Never again during its life would the jay touch a monarch butterfly. It even avoided the perfectly edible viceroy butterfly, which has developed the camouflage of closely resembling in color and pattern a monarch.

Figure 10.2. Left, Monarch caterpillar[iii]. Right, black widow spider.[iv] In both cases the bright coloration is a warning to potential predators that the animal is toxic. For greater efficiency, often different toxic animals evolve to look similar to each other (so that potential predators have to learn only once) and some animals, such as a viceroy butterfly, evolve to resemble the toxic adult monarch butterfly.

Thus selection pressure can be quite effective, and it can change species over relatively very short periods of time. Although the types of selection just mentioned relate to predation, many other forms of selection are even more effective. A disease that races through a population, killing 99% of a population, can convert the population from 99% susceptible to 100% resistant within a generation. We know that epidemics have decimated parts of the human population from time to time: plague in the Middle Ages and Renaissance killed at least 1/3 of Europe, with more in some regions. Perhaps 90% of the native populations of the Americas died shortly after Europeans arrived, from smallpox, tuberculosis, and minor respiratory diseases to which they had no resistance. In the Northeast of the United States in the early 21st C, the crow population was nearly exterminated by West Nile Virus, and the raccoon population by rabies. Many other factors can also change populations: a rapid change in weather can destroy all those that cannot resist cold, heat, or drought; or a single, pregnant white pigeon can be blown to an island, where there are no natural predators, and found a population that consists of entirely white pigeons. Climatic conditions, for instance a drought that causes plants to produce smaller, harder seeds will yield among birds that eat the seeds a strain with smaller, heavier beaks, within one or two generations. Among bacteria, introduction of a new antibiotic can lead to the death of 99.9% of the bacteria. If one in 1000 survives, however, it can continue to breed and produce a population of antibiotic-resistant bacteria within a matter of days or weeks. This is why most scientists worry about the overuse of antibiotics, particularly in low "prophylactic" doses. Such selections have been observed many times in the field and in the laboratory. The standard means of isolating specific genes or antibodies (to use as reagents or as "smart drugs") consists of manipulating bacteria so that the only ones that survive are the ones carrying the gene of interest, thus allowing selection to sort through millions of variations to find the few that are of interest. The Lederberg experiment[v] that proved the existence of genetic recombination (mating) in bacteria was inherently selection, as are the bulk of experiments in molecular biology today. There is no question that selection exists in nature and that it can be effective in changing traits of an organism. But can selection account for the change of one type of organism to another: create a new species, or derive a bird or a mammal from a reptile?

Darwin thought that all selection (which of course was the mechanism of his hypothesis) was likely to be the gradual type described above. What would change from time to time would be the conditions rather than the style of selection. A species could extend into an new territory (Darwin raised hundreds of living plants from flotsam that had drifted from the Caribbean to England along the Gulf Stream), climate could change, or a new predator or competitor could invade the region. For instance, as we know today, the green anole or American chameleon was a common sight in the U.S. Southeast through the 1950's. It is a 4- to-6 inch lizard that can change from green to brown to gray to match its surroundings. Today it is rarely seen. It has been displaced by other anoles, invasive from the Caribbean, and now exists only as a tree-dwelling species, still present, but almost never seen along the ground. This type of selective pressure could almost certainly over time create new species, but the question arose as to whether it could create new types of creatures such as (reading from the fossil record) frogs from fish-like creatures, reptiles from frogs and salamanders, and mammals and birds from reptiles, or all of the very different types of insects. The question was particularly acute since extremely few truly intermediate forms, such as an insect with primitive wings, or something intermediate between a butterfly and a beetle or fly, or a cross between a fish and a frog, could be found. To jump ahead of the story: some such intermediates have now been found, but they are quite rare, and they exist only transiently in the fossil record. Furthermore, there is now good evidence that many other, truly bizarre, versions of life existed very early in the history of our planet, and that what survives today is a small selection of that very wide range, but what was defined over 300,000,000 years ago. This realization has led to two modifications of some of the forces that drive evolution: salutatory evolution and punctuated equilibrium.

Saltatory evolution

Darwin considered that all evolutionary steps must be gradual, as he could not conceive of any mechanism capable of providing major leaps in evolution. The fact that the fossil record indicated long periods of stability for some species, such as trilobites, preceded by the disappearance of a more primitive type and followed by replacement by expansion of a more advanced type, was best explained as a failure in the archeological record. This argument was particularly aggressively pushed because earlier (and later) versions describing "hopeful monsters" tended to impute foreknowledge or direction to the course of evolution, which logically would not work: a clumsy appendage that would be a future wing would, for its bearer, simply be a clumsy appendage, against which selection would operate. Thus the question became, why are critical transitional creatures not seen, and why do new forms appear rather suddenly in the archeological record?

Today we know of some means by which rather abrupt, if limited, changes can occur as the result of a rather simple genetic change. Larval forms, which look very different from adults, can become reproductively mature and embark on careers as a non-metamorphosing version of another species. Some salamanders do this and can be forced into a metamorphosis that they now no longer naturally experience by the addition of appropriate hormones. Loss of growth hormones or response to them can create midget versions of larger species. And the much greater sophistication of body plan of, for instance, a fish as opposed to a much simpler chordate may be associated with duplication of the homeotic genes, the genes that regulate how our body plan changes from front to back or top to bottom[vi]. With extra sets of these genes, it is possible to designate a different role for each set, allowing much more subtle changes in overall body plan. In plants in particular, which cope much better than animals with extra sets of chromosomes, some species have apparently arisen by inadvertent cross-breeding or duplication of chromosomes. A relatively modest change in the rate of growth along one axis or another can turn the shape of an embryonic fish from that of an eel to that of a sunfish. See [animation][vii] However, it has not been possible to identify any such mutational change that would create a hitherto unknown organ, simultaneously giving its bearer access to a new lifestyle. Thus the idea of salutatory evolution does not garner much support today. Nevertheless, it remains true that, in the fossil record, many species appear rather abruptly, persist through many eons with little change, and then become extinct. The hypothesis as to why this should be so is described as the hypothesis of punctuated equilibrium[viii].

Punctuated equilibrium

In 1972, Niles Eldredge and Stephen Jay Gould proposed a different interpretation, which they called "punctuated equilibrium". In essence, the hypothesis stated that transitions were fast and brutal. Under rapidly changing conditions and very high selection pressure, a new form, adapted to the new conditions, could be selected for and replace an older form in a few tens of thousands of years, barely an eyeblink in evolutionary history. Once the new form was established, it would do well, changing very little (stasis), until the next severe geological event. These transitions characterized jumps from one species to another, rather than jumps from one type of animal to another. However, such changes could, in some circumstances and if repeated, lead to new types of creatures, as can be seen in the very rare but revealing finds called transitional fossils. Transitional forms have been found, but they are rare. One that created quite a stir was the Tikaalik fossil, a true, 375 million year old, creature that appears to be a nearly perfect intermediate between fish and amphibians.

Figure 10.3. The postulated appearance of the tiktaalik[ix] proto-amphibian.

Discovered in 2004 by Neil Shubin and colleagues, it is an aquatic creature capable of supporting itself on land, and it has many characteristics that identify it as a fish and many that identify it as an amphibian. It also illustrates the difficulty of finding such intermediates. The team found this fossil by deliberately searching for exposed land representing dating from this period, the Devonian. The located a land in northern Canada which, in spite of its harshness today, was at that time a warm land near the equator. There are very few such sites available. Most have been destroyed by ensuing geological processes or are now deep under water. Other transitional fossils that we know about include dinosaurs that have feathers, used as insulation as they first developed the ability to keep their bodies warm, and the first bird-like creatures that had feathers and wings, but also teeth, a reptilian tail, and a rather poor ability to fly. We can see their feathers as imprints in the mud in which they died, and we can even identify their colors (black and white stripes, with some red). We can also trace where whales came from, including recognizing when they first became deep-sea divers. Of course, if we look today, we can find many that we consider not to be ancestral (since they are around today) but similar to those that may have been transitional. These include mammals that have reptilian scales as well as a little fur (armadillos) and swimming, egg-laying, furred mammals that lay eggs and produce milk as a sort of sweat-like secretion from poorly developed mammary glands (Fig. 10.4). Marsupial animals, like kangaroos and opossums, do not have a true uterus in which the embryo develops to an advanced state. Rather, the young are born at a very early stage and crawl into the mother's pouch where they suckle until they become more capable of a more independent life. Marsupials most likely do not represent a transitional form to more modern mammals, but most likely are the descendants of another evolutionary experiment in producing a mammal-like protection and nourishment of embryos. They survived primarily in lands where they did not encounter competition from true placental mammals, and even developing astonishing diversity. There were and are marsupial equivalents of wolves, rabbits, and moles. (Fig. 10.4)

Figure 10.4. Left: Duck-billed platypus. Right: Tasmanian wolf or Tasmanian tiger. It is well worth looking at the video clip [x] to appreciate how very doglike the creature was.

Jury-rigging

The rearrangements and jury-rigging by which dog-like land carnivores were converted into whales are seen throughout evolution. For instance, the light-sensitive part of the cells in our eyes (rods and cones) face inward rather than outward, with the part of the cell that transmits the signal facing outward. This arrangement decreases our low-light sensitivity and slightly blurs the image. The octopus does it better, with the sensitive sides of the cell facing outward. The arrangement exists because of how our eyes evolved and now develop, as an outgrowth of the brain (Fig. 10.5). Another major issue is why we get backaches. To stand erect, we more or less balance our spines on our pelvis, much as one juggles a broomstick on one's hand. This constant correcting and swaying is quite tiring to the muscles that support the spine and requires us to shift the weight to something else by sitting down. Also, our spine has an S-shape curve, leading to many problems in posture and, in some elderly individuals and some diseases, finally exaggerating to the point that it compresses the rib cage, making it much harder to inflate the lungs and becoming an ultimately life-threatening condition. Why should this be so? An engineer might have designed a much more stable arrangement. However, our spines were not originally designed for an erect posture. The original vertebrates had limbs thrust out to the sides and walked with considerable side-to- side undulations, much like a swimming fish, a movement we tend to describe as crawling, and animals that move in that fashion, vermin. This waddling motion can lead to quite rapid movements but is energetically inefficient, and the position of the limbs means that the animal cannot lift its body far from the ground. (Fig. 10.6. Try doing pushups with your upper arms held parallel to the floor.)

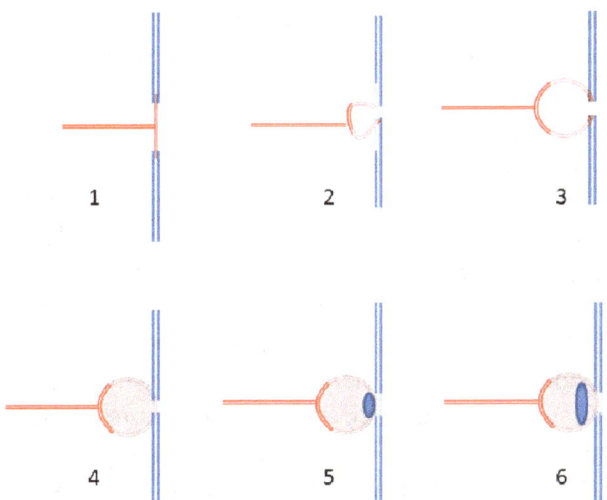

Fig. 10.5. Summary of evolution of eye as seen in a series of molluscs. The sequence is similar though harder to document in vertebrates. 1. Photosensitive cells (red) are simply neurons that reach the surface of the animal and are more sensitive than others to visible light. (Your skin cells can detect the heat generated by the light of the sun if the light is intense enough. This allows the animal to distinguish light from dark. 2. Sinking the receptor under the surface and, preferably, surrounding it with dark pigment allows the cell to distinguish the direction of the light. 3. Narrowing the aperture creates a pinhole camera, which can resolve crude images. 4. Filling the cavity with fluid helps to maintain a constant shape, and the refractive index of the fluid can be adjusted to help resolution. 5. Creation of a lens (blue) separate from the epidermis and a transparent cornea (light blue) external to it permits formation of high-resolution images. 6. Movement of the lens deeper into the eye and development of a second fluid area permits changing shape of lens to focus on different distances.

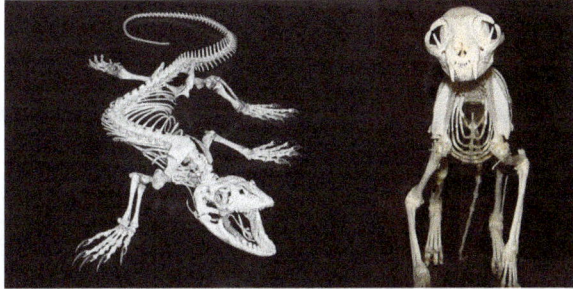

Figure 10.5. Note how in a mammal (cat) the legs are directly under the shoulders and support them, while in a lizard (right) the legs spread to the side and only with difficulty can lift the body[xi]

About 275 million years ago one group of reptiles, known as the therapsids, evolved a different configuration, with the upper arms and legs perpendicular to the ground, so that all movement of the limbs was anterior to posterior (front to back) and the body could be held straight during movement. This form of movement was far more efficient and meant that the body could be lifted higher off the ground. It became the dominant means of movement among the mammals, which derived from the therapsids. Spines evolved accordingly. They became slings from which the organs were hung, supported by limbs at the four corners. The spine was therefore arched or, more rarely, hung between the supports. Such an arrangement does not work well, either by balance or distribution of load, if an animal attempts to stand on its hind legs. Picture the awkwardness with which a dog or cat attempts to walk upright. However, evolution does not say, "Hey, we have a new arrangement. Let's redesign the spine to make it work for a two-legged creature." We work with what we have. In this case, for humans, the solution is to create an opposing curve so that the weight is at least balanced on top of the pelvis. (See Fig. 5.7) The locations and angles at which the spine meets the pelvis and skull, as well as the angle at which the hind legs meet the pelvis, have to be changed as well. The rearrangement of the spine, however, does not substantially improve stability, but evolution works with what we have, not with an ideal, planned design. One might consider, for instance, a broader base for the spine on the pelvis or a more equitable distribution of weight around the spine. (Of course, it helps to be able to see what is coming at vulnerable organs. So there is some purpose in having the organs facing front, but one might find other means of balancing the load. For instance, many water birds such as pelicans and herons fold their long necks back over their bodies as they fly, the better to balance their load in flight.)[xii]

~~~~

Chapter 11: What is a human?

This question may seem relatively obvious or useless today, although various societies, promoting slavery, racism, or seizure of resources, have found ways to exclude one or another group from membership in the category of human. To a biologist, we are a very variable species, varying widely in total size, skin color, configuration of hair, and shape of skull and body proportions. The wide variation is characteristic of domesticated animals and plants whose variations are no longer limited by predators. Nevertheless, we are all mostly hairless apes with hair limited to specific regions and, on the head, growing to great length; we capable of using elaborate speech, forming societies, creating art, and of building complex tools. Unlike chimpanzees and bonobos, with whom we share more than 98% of our DNA, we are unequivocally erect and bipedal, it takes us twice as long to reach puberty, and, perhaps as indicators of our intense socialization and interest in the motives of others, we have light-colored palms and soles, perhaps designed for signaling. Unlike almost all other creatures, we can see the whites of our eyes and therefore judge where someone is gazing or how likely he or she is to be truthful. Human babies spontaneously point to objects they want, whereas other animals must learn what pointing means (try to teach a dog that you are pointing at something), and human babies spontaneously learn to speak and, if deaf, can create their own languages. Men are 8-15% larger (weight) than women, whereas in most apes the males may be double or even more the size of females. The most important criterion that we all belong to the species *Homo sapiens sapiens* is that we can interbreed freely and that all of our young, even those derived from the most diverse parents imaginable, are healthy, grow well, and are fertile.

Humans vs apes

The question becomes more complicated as we move farther afield or delve into history. We may differ as to the rights and protections that an ape such as a chimpanzee should be accorded, but we have little difficulty in understanding that it is not human. But what about the fossils that we encounter?[i] We can evaluate them by many criteria: Did they walk fully upright? How large were their brains? What types of tools did they use? Can we determine what size colony they preferred? Did they build structures in which to dwell? Did they wear clothing? Did they use fire? Did they domesticate animals or plants? Did they have musical instruments? Did they have a sophisticated language? And, most importantly, did they leave any kind of artifact (artwork, statues, symbolic tools) that suggests that they thought about who they were and how they related to the earth? Did they care for wounded, deformed, or weak members of their society? Did they bury their dead or leave any indication that they had a sense of afterlife or a religion?

These questions have meaning when we consider our ancestors and most especially the ancestor that immediately preceded and to some extent overlapped with us. Current evidence indicates that the Neanderthal people—we will use that term—contributed very little to the population of humans that now covers the earth. Their DNA, insofar as it has been successfully analyzed, is too different from ours. And yet they met all or most of the criteria mentioned above. (There is still dispute as to whether or not they could have had a clear language.) In brief: they made tools, they buried their dead, perhaps with some ceremony, they could control fire, they cared for their wounded, they decorated their bodies, and they apparently had musical instruments. And yet they did not survive. Therefore the question becomes, what is the species we call *Homo sapiens sapiens*? Where did we come from, and how did we come to populate the world, as opposed to any other species similar to us or not? For this kind of analysis we look primarily to the fossil record, with some cross-referencing of our ability to interpret the record in our genes. For clarity, we can use the following terms: *anthropoid* or human-ish: tailless apes that can stand erect on occasion; *hominid* (human-type): truly erect creatures with brain size larger than apes; *human*: truly erect large-brained creatures with sophisticated tool-making capability, the ability to control fire, and signs of culture. Our story begins approximately 4 million years ago (if one starts with the earliest creatures that resembled humans) 1,600,000 years ago (if one starts with creatures that were sufficiently like us to be considered within the genus Homo) or 160,000 years ago, if one considers those similar enough to us to be considered modern *Homo sapiens* with fragments of skeletons found in eastern Africa. Because the skeletons are very fragmentary, much of what we understand about their lifestyles is inferential. In general, skulls or parts of skulls are more frequently preserved than other bones. Apes have sharp, tearing canine-like teeth, while modern humans have a mixture of grinding and more gently tearing teeth. The size of the mandibles is also meaningful, since apes have more massive jaws. Although the intelligence of individuals does not correlate with brain size, in general populations of animals with larger brains are smarter than populations with smaller brains, and brain size has expanded very rapidly in human evolution (Fig. 11.1). The vertebral column of humans follows an S-shaped curve, to balance the torso on the pelvis, and humans have flat walking feet unlike the prehensile (grasping) feet of many apes (Fig. 11.2). Originally scientists asked when humans became fully erect, because some hypotheses considered that the increase in intelligence followed the free use of hands by our ancestors; but unfortunately vertebrae and feet were only rarely found. In fact, it was typical of earlier dioramas, or model displays of early human existence, that these individuals were shown in tall grass so that the shape of their feet would remain ambiguous. Today we recognize clues that can provide indirect evidence. In truly erect humans, the skull must be balanced on the spine, with weight toward the back approximating the weight in the front, and the line of the eyes is at right angles to the spine. An ape's head sits at more of an angle to the skull. Thus the angle of attachment of the spine to the skull can be interpreted (Fig. 11.3). In a similar manner, the femur (thigh bone) of a four-legged animal rests naturally in the pelvis at approximately right angles to the spine, whereas in humans the femurs are almost aligned with the spine (Fig. 5.7). In the pelvis of an ape, which tends to "knuckle-walk" (move without putting its full weight on its front limbs, but using them for balance and to shift weight) the alignment of the pelvis is at an angle. In humans the pelvis is rotated relative to that of other animals to act as a basin for supporting the viscera. If a femur or a pelvis can be located, we can infer the posture of its owner. More recently, higher-resolution and reconstruction techniques have led to further inferences. For instance, an ape that climbs and swings from branches with its arms needs room for strong shoulder and forearm muscles. It therefore has a narrower, barrel-shaped chest than does a human, adapted for longer-endurance running and not particularly powerful arms. Neanderthals left Africa and spread into Europe, going eastward as far as Siberia, but did not reach the New World or, with few exceptions, the warmer regions of Asia. A genetic subset of modern humans, probably a few small bands or tribes, left Africa 70-50,000 years ago and, mostly following coastlines, spread throughout the world very rapidly for a species but basically at the rate that would be matched if each generation wanted to move out of range (a mile or two) of their neighbors (Fig. 11.4)

Figure 11.1. Early humanoids. From left to right: Australopithecus; Homo ergaster; H. habilis; H. heidelbergensis; H. neanderthalensis as portrayed in the Cradle of Mankind Museum, Sterkfontaine, South Africa. Hair shape, color, and distribution and skin color are speculative, though there is evidence that Neanderthals carried genes for red hair. Human distribution of head hair (not beard) probably was an adaptation to upright locomotion in strong sunlight, to reduce heating of the brain, whereas hairlessness was probably established much more recently, perhaps only in modern humans.

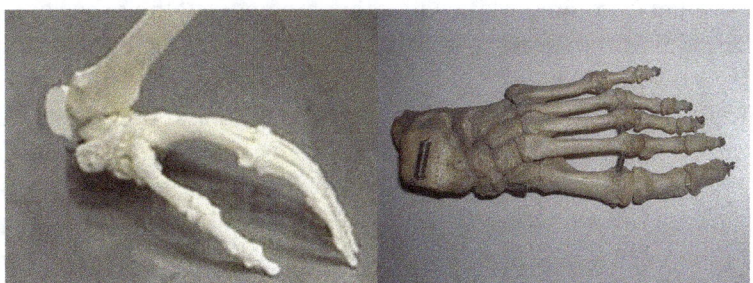

Figure 11.2 Chimpanzee (left) and human (right) feet. The arched construction of the chimpanzee foot and the greater separation of the big toe from the other toes allows the chimp to climb more easily, but makes it more difficult for it to walk. Its wider stance and angled articulation of spine to pelvis also indicate that it is a relatively poor walker.

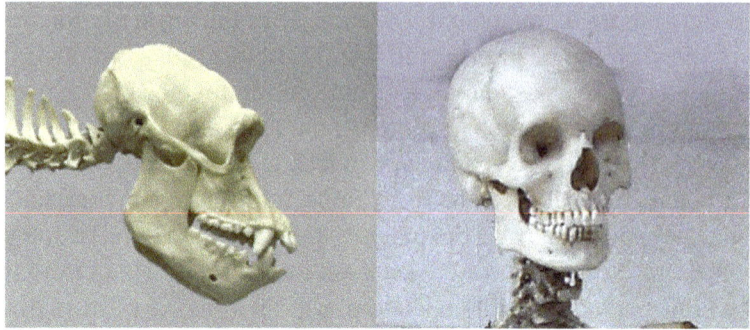

Figure 11.3 Chimpanzee (left) and human skulls. The articulation between the spine and the skull, and the weight distribution of the skull when the eyes look forward, indicate that clearly the chimpanzee does not normally stand erect, whereas a human does.

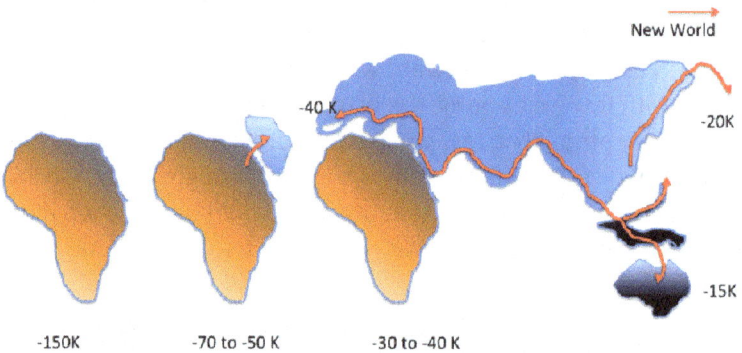

Figure 11.4. Human migration, as determined by genetic diversity and traces of human activities. Genetic diversity. The colors indicate different primary genetic markers. Orange = substantial diversity; blue to black = far less diversity. Dates = years before present era. *Arrows:* Migration patterns as traced by earliest evidence of human activity. To the right, humans reached the northern New World approximagely 20,000 years ago and the southern tip of South America approximately 15,000 years ago. In general, blood types, genetic evidence, and general appearance (skin tone, hair color, hair form, extent of beard and hirsuteness) are consistent with these values. Credits: Shreeve, James, 2006, The Greatest Journey, National Geographic, March 2006

THE IMPORTANCE OF LANGUAGE

Although all animals communicate with other members of their species, and some social animals cooperate to hunt, forage, or migrate (listen to migratory geese passing overhead) true language is much more powerful and much rarer. While porpoises seem to have a quite complex communication system, human language may be truly unique, and it is clearly much more efficient, creating the possibility of a species improving its probability of success by a means other than mutation. Consider, for instance, a pack of chimpanzees or wolves trying to encircle an animal that they hope to trap. Grunts or varying sounds and gestures may work well, but "Bill and Fred, go hide behind those bushes. Mike and I will drive the antelope toward you. When it gets to that rock, you should be able to spear it," is far more effective.

Recent history has given us an example of how selection for intelligence— here, the ability to speak, address subtle concepts, and remember—might have functioned. Just after Christmas 2004, a severe earthquake off the coast of Indonesia generated a huge tsunami ("tidal wave") in the Indian Ocean that devastated the coastline of Indonesia, India, Sri Lanka, Thailand, and other countries and resulted in well over 170,000 deaths. During the cleanup, rescuers assumed that the Moken people of the Andaman Islands, some low-lying islands directly in the path of the tsunami, had been lost. They were very startled to find that, although the villages were completely destroyed, most of the people had survived. What had happened was the following: First, the people of the islands derive from migrants who originated in southern Africa and who presumably populated the southern Indian Ocean coastlines, Melanesia, and Australia, where the descendants became the people known as the Melanesians and the Australian aborigines. They lead a simple life as fishermen in small villages, presumably similar to early human societies: they do not concern themselves with time, have no means of expressing how old individuals are in years, and have no words reflecting future and past. In their village life, they recount ancestral tales.

The water in a tsunami must come from somewhere and, because of the physics of wave motion in water, it pulls the water in from in front of the wave (Fig. 11.5). Thus a tsunami is preceded by what appears to be an extraordinarily low tide, most likely occurring at an unexpected time. For a tsunami of this size, generating waves up to 60 or 80 feet in height, the water withdrawal was enormous, and preceded the arrival of the wave by up to one hour. Many of the people who died in the tsunami had been intrigued by the surprising tide and had gone out to the suddenly-exposed beachfront to gather shellfish and stranded fish. Not so the people of the Andamans. When they saw the water retreat, they ran for the hills, and their fishermen who were at sea headed for the deepest water they could reach. Burmese fishermen continued collecting squid and were lost, since the height of the tsunami is built in shallow water. (Watch how waves come in from the ocean: they get taller and finally break as their bottoms scrape the sand.)

Figure 11z. Mechanism of a tsunami. As in Hokusai's classic illustration[ii], a wave is not a mass of water moving forward. Rather, it pulls in water from in front of it, and the water cycles through the wave and back under it. When a wave approaches the shore, the drag along the bottom slows the bottom relative to the top, and the wave builds height and finally falls over. Similarly, a tsunami draws the massive amount of water that it will take from in front of the wave, causing an unusually low tide or retreat of water from the coast, before it arrives See also argument why tsunami builds in height as it approaches shore.[iii]

Why did the Andaman people run? The tales that they told of their ancestors included stories of when the sea suddenly ran away and then came running back. It mattered little that they felt that the sea, responding to the spirits of their ancestors, was angry and came to eat the villages; the description in their legends was accurate, including the withdrawal of the sea and a first, smaller wave followed by a second, larger one. Geological records indicate that tsunamis had most recently occurred 40 and 200 years before, but the collective memory, explicitly described (as opposed to grunts) informed the people and allowed them to survive. The Andaman people lived very close to nature and observed it very well. Their comment about the Burmese fishermen was that the Burmese "were just looking at the squid, and were not looking at the water." Picture an earlier time, when small groups of the first modern humans were spread widely in the region, and the total population was small. In this situation certainly and most likely in many similar situations, the only survivors would have been that group that was able to profit from the memory of the earlier experience. There are two elements here: good memory (which is also shared by some animals) and above all, ability to communicate that memory from one generation to another (possessed by very few if any other animals) would have been an extremely influential factor in the evolution of the human race. This would be particularly important if it concerned ways of hunting, making tools or boats, or surviving difficult times. Many linguists feel that the first human languages were similar to the surviving click languages of the Khoisan and other peoples of Africa. These languages have very peculiar clicks[iv] in the middle of syllables[v]. Some have noted that animals that humans hunt are much less frightened by clicks than by sounds made by human vocal cords and they therefore suggest that the first languages were used for hunting. It is probably quite meaningful that one human gene that has evolved quite rapidly and is very different from that of chimpanzees is a gene that is very important for the understanding and use of speech. DNA extracted from Neanderthals indicates that they too had the gene, but other evidence suggests that they had not learned to speak. The rate of improvement of stone tools increased dramatically approximately 100,000 years ago. Since anatomically modern *Homo sapiens* first appeared 200,000-160,000 years ago, this change in rate may reflect the mastery of true language.

It is reported that some animals also fled to the hills. We are not aware that these animals were capable of communicating from one generation to another, for in almost all species, adults do not interact with younger animals beyond the earliest phases of child-rearing. It is possible, however, that they could detect sounds lower or higher than humans can detect, or of lower intensity, and thus heard the incoming tsunami. They might have also detected other changes, such as in light or in magnetic fields; or they may simply have panicked in the face of a rapidly changing and unknown situation, not being as curious or self-assured as modern humans. At this point we do not know.

THE ORIGIN OF THE HUMAN SPECIES

There are several lines of evidence suggesting the origin of humans. The oldest and most well-known evidence is that of fossil skeletons, or rather pieces of skeletons. These have been known, and in increasing numbers, since the mid-19th C. Other evidence includes traces of human activity in various parts of the world, coupled with radiometric and geologic dating; examination of the physical appearance of different groups of humans; consideration of the different versions of human language; and, today, several means of analyzing and comparing DNA from humans around the world. As has been discussed before, the evidence converges on common dates, and the convergence is a major basis for confidence in the hypothesis.

The physiology of humans and that of apes is extremely similar, as is the biochemistry. In fact, over 98% of human DNA is identical to that of the chimpanzee and slightly more to that of the pygmy chimpanzee or bonobo. To understand human evolution, we must first agree on how humans and apes differ and, hopefully, identify means by which those differences can be detected. The obvious differences include the size of the brain in humans, which is approximately three times that of chimpanzees; the relative hairlessness of humans, which cannot be measured in the fossil record (but can be surmised by the evolution of lice); the fact that humans are truly erect and bipedal, whereas apes are not; humans are much more omnivorous than apes, which tend to eat more fruit and less meat, and their dentition (shape of teeth) is correspondingly different; and the skulls of apes differ in that they have pronounced brow ridges, slanting foreheads, a flat nose, massive mandibles and prognathous jaws (with the teeth jutting well in front of the nose). In terms of soft tissue, apes do not have lips (outwardly turned tissues in the mouth), and apes cannot talk. This latter point indicates soft tissue changes, since the much lower position of the human larynx permits the complex changes in shape of the throat muscles to generate all the sounds that humans make. Apes cannot make the several dozen individual sounds that humans can, but on the other hand the lowered larynx makes it possible for humans to choke on food. Unfortunately soft tissue like the larynx is not preserved in fossils, though several anthropologists have attempted to infer from the structure of skulls (indications of where jaw muscles inserted) whether or not any of the fossils were able to talk.

There are many more subtle consequences of the differences in lifestyle between and humans and apes. To walk easily and efficiently on two legs requires, for balance and conservation of energy, longer legs. The short legs of the chimpanzee and, because of the orientation of its pelvis, its wider stance, give it a waddling, inefficient gait compared to the straight front-to-back thrust of our walking movements. Arms of animals that can swing from trees require strong muscles, leading to a chest that is narrower at the top (to allow for the muscles) than the bottom. Even the spines on the vertebrae serve for attachment of back muscles, giving indication of the size of, and demands on, these muscles. Thus it is possible to extrapolate from fragments of bones to an image of the creature from whence the bones came.

There are also indications of intelligence that, at a higher level, may be recognized. Apes can use primitive tools: Orangutans can strip leaves from twigs and use the resulting sticks as probes to retrieve insects from nests; the preparation of the twig indicates foresight and deliberate intent. Chimpanzees can use rocks to break open large seeds or to throw at a predator or other threat, and they may even protect their feet with shoes of leaves if they walk over thorns or other sharp terrain, but none of these uses leaves identifiable remnants. Early humans learned to split rocks to get sharp edges. Later they learned to hit specific types of rocks to create thin, sharp blades and even to sharpen the blade when it got dull. The use of tools to scrape meat from bones can be identified from the marks left on the bones. Even later humans used fire to create metal tools. All of these activities leave remnants that can be identified, and fire leaves its own traces. More sophisticated tools such as throwing tools (spears, javelins) and fishhooks can be identified by their shape. Creatures that can control fire also have access to far wider ranges of food, since many plant and animal products are indigestible to humans unless they are cooked. Some of the diet may be inferred from the shells and scraps left by the eaters—early humans had a rather sloppy lifestyle, resulting in the collection of middens, or trash heaps, that can be explored—and indigestible materials in fossilized feces give further clues. Finally, art—organized markings, even if their purpose is not understood[vi], as well as jewelry—is characteristic of all truly modern humans. Anthropologists and evolutionists delve through all of these cues to trace the origin of humans (see for instance the artwork of the Lascaux and Altamira caves).

Primates have existed for approximately 60 million years. The least evolved primates are small tree-dwelling creatures called lemurs, pottos, lorises, and tarsiers. They are very cute animals because, unlike the anthropomorphized stuffed bear and other toys, they truly have forward-facing eyes and opposable thumbs (thumbs that can touch the opposite side of their paws, or hands). Both of these characteristics are adaptations to scampering through trees, the forward- facing eyes giving binocular vision so that the animal can judge distance and the opposable thumb and big toe allowing the animals to grasp branches. Likewise, unlike many other mammals, primates rely more on vision than on smell, and they have color vision. All primates retain these characteristics. The so-called New World monkeys are similar but are monkeys (anthropoids, with more human-like body shapes and faces) but have an interesting adaptation, in that many have tails that can be controlled and used to hold onto branches (prehensile tails). Monkeys like these have been around for about 35 million years.

A little over 20 million years ago, in Africa, arose a different version of monkey, one which was also at ease on the ground, could stand and sometimes run on two legs, and whose tail might serve for balance but was not effective for grasping and climbing. All of these monkeys are very social and move around during the day as opposed to night. These are the so-called Old World monkeys. About 15 million years ago, the true apes or hominoids arose. These mostly large monkeys are tailless, and although two apes, the orangutan and gibbon, are almost entirely arboreal, the others, which also include chimpanzees, bonobos (pigmy chimpanzees), and gorillas, are reasonably comfortable on the ground. These "higher" primates have another characteristic that is relevant to the story of the evolution of humans. Almost all female mammals have "estrus" cycles. In an estrus cycle—the word derives from the Latin for "excitement"—the female accepts advances of a male only when she is ovulating, and otherwise actively drives him away, and after ovulation her ovaries do not produce progesterone (an early hormone of pregnancy) unless she has mated and is likely to become pregnant. If she does not become pregnant, she sheds the ovulated eggs in an increased secretion that quantitatively does not approach a menstrual cycle. Female higher primates have menstrual cycles. In a menstrual cycle the female may accept the advances of a male at any time, and after ovulation the ovary routinely produces progesterone for about two weeks after ovulation, preparing the uterus for a potential pregnancy. If pregnancy does not ensue, the tissues in the uterus break down and are released, producing a true "period". These latter characteristics of course apply to humans, linking humans to the higher primates. The continuous sexual receptivity of the higher primate female plays an important behavioral role in keeping the males in proximity, allowing pairing or, minimally, the organization of troops in which the babies are protected. When did humans become distinct from apes, and what evolutionary mechanisms led to the appearance of humans?

The close similarity of human to chimpanzee/bonobo DNA does not mean that we are the same. It is important to understand that these DNA sequences are measured on current, living, animals, not ancestral animals. Also, genes can differ by being turned on and off at different times, as well as by producing different proteins. Since humans in many respects appear to be apes that are sexually mature while retaining many characteristics of infants (prolonged growth of legs, hairlessness, curiosity and capacity for play), much of the difference between humans and apes may reside in the timing of when genes become active. Biochemically we may not be very different but, if we are the equivalent of sexually active juveniles, our behaviors may be very different.

About 30 million years ago, the great forests of eastern Africa were beginning to dry up. India was pushing its way into Asia, and the uplift of the Himalayas in addition to the appearance of a new landmass in the area changed the weather pattern. The forest gave way to an expanding savannah. By 6 to 7 million years ago, the savannahs were fully established. An ape that could easily move across grasslands would be able to move from one wood to another, and otherwise expand beyond the forests. Thus at about this period we find a type of skull that differs from that of most apes in interesting ways. The skulls have smaller canine, or tearing, teeth, and their faces may have been a bit flatter. The position of the foramen magnum, the hole at the base of the skull where the spinal cord enters the skull, is a bit farther forward than in apes, suggesting that the skull is more balanced on the spine, meaning that the creature is more comfortably balanced in a vertical position. Thus what defines these creatures as being related to the human line is the suggestion that they were more fully bipedal than other apes and that they were eating a more varied diet. Unfortunately, the feet, which are grasping with opposable big toes in apes and flat in humans, are typically lost in the fossil record.

The story begins to get much more interesting about 4 million years ago. Reliable dating of the soils in which they are found indicates that various members of the genus named *Australopithecus* (Southern ape) lived between 4 and 2 million years ago. These creatures were fully bipedal and had teeth much more similar to those of humans. Their brains were still relatively small, about the size of an orange. This is equivalent to the brain of a chimpanzee, which is of similar size to these creatures. "Lucy," a 40% complete skeleton found by the Leaky family in Ethiopia, was an *Australopithecus* who lived 3¼ million years ago. Her arms were still relatively long, suggesting that she could easily climb trees, but the pelvis and the remnants of the skull argue for an upright posture and bipedal locomotion. Even more convincing was the discovery of 3.5 million year old bipedal footprints in what is now Tanzania. These footprints were made as a presumptive *Australopithecus* walked across newly-fallen ash from a nearby volcano. The ash subsequently was wet, probably by rain that was "seeded" by the ash in the atmosphere, and solidified into rock. Even more convincing and touching is a detail about these footprints: They start out as two sets, an adult and a child. About half way along, the adult picked up the child, as evidenced by the disappearance of the child's footprints and the deeper impressions made by the prints of the adult.

At approximately this time we encounter some very important fossils, not of humans but of animals found in Ethiopia. What is interesting about these fossil bones, from about 2 ½ million years ago, is that they have marks on them that indicate that the flesh was scraped from the bones by stone tools. This is the first evidence for tool use. Thus *Australopithecus* appears to have been a bipedal, upright, tool-using creature.

Between 2.5 and 1.6 million years ago, fossils appear that are similar enough to humans to warrant the genus designation *Homo* (man, or [the] same [as us]). How do they differ from the *Australopithecus* type? They have skulls that can accommodate larger brains, about half the size of modern humans. Associated with these skeletons are well-made and sharp stone tools, enough that the first of these type of fossils were given the name *Homo habilis* ("handy man"). Meanwhile, other hominids persisted along the *Australopithecus* line, continuing with small brains and prognathous (jaw forward) face. This was previously a considerable source of confusion, as long as people imagined a direct lineage from the most primitive anthropoids directly to modern humans. Today we recognize that, similar to other sequences of evolution, there were many branches to the line that led to humans, most of which finally petered out. Thus, contemporaneous with *Homo* were several other hominoids with one or more characteristics approaching those of modern humans, but these were not part of our ancestral line and ultimately the lines died out.

The *Homo* line gave rise, between 1.9 and 1.6 million years ago, to a very interesting new variant, named *Homo ergaster* (working man). This creature now had a brain 70% of modern size (900 cc compared to 1300 cc). Skeletons of *H. ergaster* indicated tall, long-legged individuals with hips clearly structured for straightforward, long strides. Perhaps related to their ability to walk long distances, skeletons of *H. ergaster* are found over much wider areas and in more arid lands. Their teeth were of a more generalized style, indicating a wider variety of food. Their fingers were too short and straight for them to have been good climbers. Their stone tools were sharpened with skill. Unlike the male/female size ratio of 1.5 of *Australopithecus*, in *H. ergaster* the ratio is 1.35, much closer to the ratio for modern humans of 1.08-1.2 (about 5% difference in height, 20% in weight; only height can be reliably measured in fossils). The decreasing ratio suggests less male-male competition for females and therefore more pairing of partners. (Logically, it would seem more sensible to have females bigger than males, but in the way that the world is constructed, males frequently fight over females and, not only does the larger male often win, the female often prefers the larger male. Thus, where there is competition among males, there is heavy selection in favor of increased size of males.) One driving force may have been a longer period required for infant care. (Human babies take twice as long to reach puberty, and therefore twice as long to increase their knowledge before becoming independent as do chimpanzees. This slowing of maturation is important and is probably related to the relatively juvenile appearance of humans as apes, and may also be part of the process that gradually lengthened human lifespan to approximately double that of apes.)

A more recent version of the genus *Homo*, *Homo erectus*, is considered by some to be simply a late version of *H. ergaster* and by others to be a different species. It may be nearly a semantic argument, since *H. erectus* survived into much more recent times—from 1.8 million years ago to as recently as 200,000 years ago, but *H. erectus* continued the trend and was the first hominoid to leave Africa. Remains of *H. erectus* have been found in the Republic of Georgia, and in Indonesia.

Finally, in 1856 a most curious hominid skeleton was found in a cave in the valley of the Neander River in Germany. Named the Neander Valley, or Neanderthal, skeleton, it was very human in many respects. For instance, it had a brain size equal to or exceeding that of modern humans (approximately 1300 cc). On the other hand, the Neanderthal people were much more heavily boned than modern humans; they had heavy, massive jaws; and they had pronounced brow ridges (Fig. 11.1; this is a rather generous portrayal; many presume much grosser features; see also Fig. 11.7). Their hip sockets were a bit different from ours, indicating that they walked with a more waddling gait. Everything about their skeletons (strength of bones, places of attachment of muscles; shape of thorax providing room for much larger pectoral muscles and of skull, providing room for much more powerful jaws) indicates that they were far more muscularvii than modern humans.

However, based on the material found among them, they did many things that were essentially human in style. Since the period in which they existed, 200,000 to 40,000 years ago, was an ice age in Europe, they had to use and control fire, and it is difficult to imagine how they could have coped with the winters unless they had clothing. There is evidence that they constructed wooden homes, or nests if you prefer, on platforms beside lakes in Switzerland. Some of their dead have what appears to be jewelry or other indications of planned burial. They honed stones into very effective hand axes.

However, what may be missing is definitive evidence of artwork. Although for at least one cave, modern results suggest otherwise, for the most part these people left no statuettes, drawings, or markings on stones to suggest that their thoughts surpassed the immediate and the practical. One statue, dated to Neanderthal times because it was found between two accurately dated volcanic layers, is subject to dispute both as to its age and whether it is truly a carved artifact (Fig. 11.6). Because of these lacks, we cannot be certain that we know or recognize these people. If subsequent research for the Grotte des Fées (Grotto of the Fairies) in southern France confirms the recent findings, we will have to reassess this judgment and reevaluate the issue of why these people disappeared approximately 40,000 years ago.

Fig. 11.6. One of the very few potentially Neanderthal artifacts. It may be similar to the Venuses of modern humans, but both the dating and the resemblance have been disputed. Currently available at http://donsmaps.com/ukrainevenus.html

Why they disappeared is a matter of some debate. They were apparently in some decline by the time modern humans arrived. They appeared to live primarily from the large game animals (mammoths, the gigantic Irish elk), which were disappearing as the glaciers retreated. Their diet does not appear to have been as flexible and adaptable as that of the more gracile modern humans. There is no evidence from their bones that modern humans killed them, and no evidence such as large numbers of dead that modern humans brought new diseases. Judging from the slowness in the evolution of their tools and the lack of artwork, it is doubtful that they had a language sufficiently polished for speech and cultural transmission of ideas and skills. (Anatomically modern humans existed for at least 50,000 years before their tool making "took off" in the sense that new and more efficient modifications began to appear very rapidly. From the genetics, it appears that there was very little cross-breeding between modern humans and Neanderthals, and there is still argument that the few traces of Neanderthal genes found among Europeans represent common ancestral genes rather than cross-breeding. Certainly if modern humans had language and they did not, sexual attraction, as opposed to rape, would have been limited.

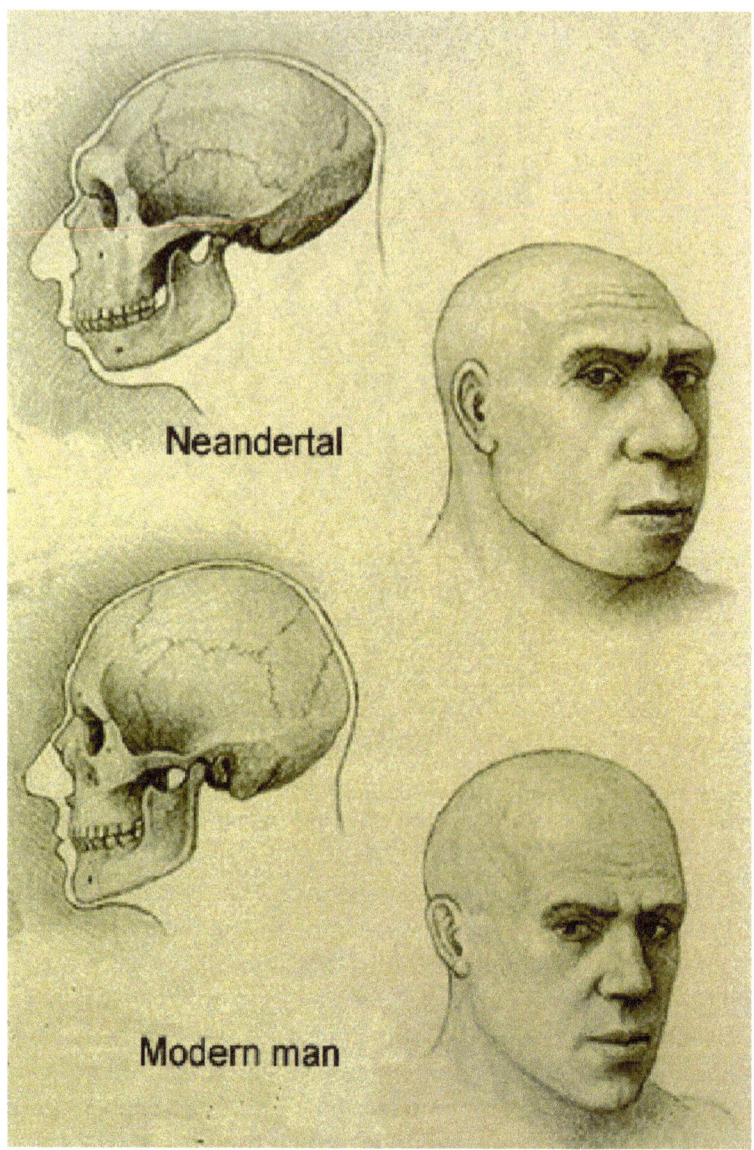

Fig. 11.7. Interpretation of appearance of *Homo neanderthalensis* compared to *Homo sapiens*. Skin color and relative level of hirsuteness are of course totally speculative. Such images help us to understand that, though the similarity of these people would have been enough to confuse us, the differences were such that they would not have moved in our societies without evoking a double-take. Their ability to speak is also in question. Credit: http://sapphire.indstate.edu/~ramanank/heads-sk.gif (no longer available)

Late in the period of the Neanderthals, starting according to DNA approximately 200,000 years ago, a different variety of hominid appeared in Africa. Other than the dry Rift Valley of Ethiopia, the African continent is not very conducive to preserving the remains of hominids, and we have little physical evidence of what was going on. What we do know is that, approximately 100,000 years ago, a new type of hominid appeared in the Middle East. This hominid had the full modern brain capacity of 1200–1300 cc, nearly absent brow ridges and a high brow, modern, multi-purpose teeth, a flattened face that had receded behind the nose; and its skeleton was lighter weight and more delicate than that of the Neanderthals. In short, this was a modern human who would not stand out in a crowd today. We do not know the colors of their skins or the form of their hair, though we might speculate, but they were modern humans, at least anatomically. Though the first exploration as far as Israel does not seem to have survived, by 70,000 years ago a second wave again reached the Middle East. Their tools were far superior to those of the Neanderthals, in that they used one rock to shape another and, rather than simply cracking a rock to get a sharp edge, by 50,000 years ago they had learned to break off thin blades, suitable for arrowheads and knives, from flint. Of considerable interest is the fact that, though for greater than one million years stone tools evolved very slowly, beginning with what we can now call *Homo sapiens sapiens* tool production began to be a specialized art and tools rapidly became more efficient and better designed, including the invention of tools that are clearly recognized as arrowheads and javelins. The sudden appearance of rapid development may reflect true language, and therefore the ability to propagate and extend culture. Most importantly, these people left traces of their passage in the form of odd geometric patterns carved onto bone and stone, curious figurines often in the shape of obese, possibly pregnant, women termed Venuses (Fig 11.8), and paintings on the walls of the caves that they inhabited. When they buried their dead, they surrounded them with tools, stones or shells that appear to have been necklace jewelry, and red-colored earth that may have been symbolic or may have been used to make them more alive-looking. Though we do not know what they intended by the carvings and the art, it is clear that the important word is "intended": these people thought and could formulate concepts, almost certainly with a sense of future and past. They understood the concept of a symbol, and they may have attempted to control their fate with the use of these symbols. They were modern humans and, though their presumptions were undoubtedly very different from ours, we could have understood how they reasoned. Although most of the caves are now closed to the public because visitors bring in new microorganisms and the lights allow algae to grow, several websites offer innovative tours that are well worth a visit. (see Further References, footnote 10 at end of chapter)

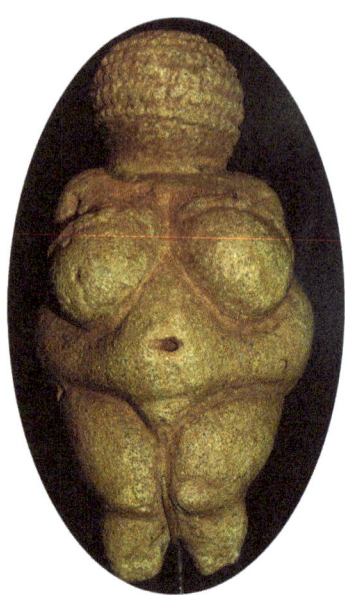

Fig. 11.8. The Venus of Willendorf. This is clearly a human female figure, wearing a braided hat and some decorations on her breasts. This was made by modern humans approximately 25,000 years ago. Compare this to Fig. 11.x

These people moved quickly around the world. By 50,000 years ago, the first humans had reached Australia, presumably traveling along the coasts past India and island-hopping thereafter—obviously by boat. They also migrated toward the northeast and northwest, appearing in both Europe and central Asia between 40,000 and 30,000 years ago, crossing the Bering Strait during an ice age when there were land bridges and, apparently following the coastlines, reaching southern South America approximately 15,000 years ago. Movement inland in the Americas was a bit slower, so that the earliest non-coastal settlements date from 19,000 to 12,000 years ago. All of these people left skeletons, burial grounds, campsites, weapons, and artwork along their trails, allowing a fairly accurate and well-confirmed record.

Though for a species this speed of expansion is quite remarkable, in terms of what we see in current human behavior it was not a headlong rush. This expansion could have been accomplished if, in each generation, the outermost family decided to move out of sight, out of earshot, or out of the hunting range of its neighbors. One or two miles per generation would have been adequate. What the movement does suggest is a desire to seek more game or more resources or to keep away from perhaps competing tribes, together with a resourceful life style that was able to overcome geographical and geological barriers. We know that these people were resourceful and that they hunted resources; in some areas of Europe there is evidence that they could dig substantial distances into the ground to locate good flint for arrowheads.

There are many fascinating questions here. Among these questions of course are, what happened to the Neanderthals, where modern humans came from, how we know their origins, and how the world ended up with different races. The Neanderthal (*H. neanderthalensis*) population had been declining, but the Neanderthals disappeared from Europe at approximately the time that the Cro-Magnon population (*H. sapiens*, named from the cave in France from which they were first clearly distinguished from *H. neanderthalensis*) spread through Europe. Although skeletons of *H. neanderthalensis* do not display wounds indicating attack by humans, the question of whether *H. sapiens* fought with them, made them uncomfortable enough to cause them to retreat, carried new diseases to which Neanderthals were susceptible, or interbred with them, ultimately diluting them out of existence, is of interest to scientists and most amateurs. It is also possible that the Neanderthals, who depending on how one counts the variants of *H. erectus* survived between almost two million years and a few hundred thousand years, could not adapt readily to the changing conditions at the end of the last ice age. From their skeletons, tools, and middens, and from variants of radioisotope dating, it seems clear that they were meat eaters and hunters of big game such as mastodons and Irish elk, which were dying out, while modern humans were more willing to experiment with small animals, fish, and grains. The last Neanderthals disappeared from the southern Iberian peninsula perhaps as late as 40,000 years ago, in which case they would have encountered modern humans. However, there are corrections to be applied to radiodating because of diets (there are differences in $^{14}C/^{12}C$ between marine and freshwater fish and between fish and land animals), and recent studies suggest that the Neanderthals may have disappeared at an earlier date.

The lack of evidence of battles between *H. neanderthalensis* and *H. sapiens* led most researchers to favor the hypotheses of displacement or interbreeding, but advancing technology gave a much clearer answer. It became possible to collect DNA from first one, then a second specimen of *H. neanderthalensis* and compare the DNA to that of modern humans. More specifically, it was mitochondrial DNA that was analyzed. Sperm mitochondrial DNA is lost at fertilization, and mitochondria are inherited only from the mother. Since mitochondrial DNA is not subject to rearrangement during mitosis, it is more conserved than nuclear DNA and, since there is only one nucleus but hundreds of mitochondria in each cell, there is more mitochondrial DNA available.

If we assume that changes in some DNA bases will not markedly influence evolution ("neutral mutations") and that the probability of these mutations appearing is random, one can estimate both the evolutionary distance between two subjects and the age of their last common ancestor. From the specific sequences, one can suggest a probable lineage. From the sequences established from the *neanderthalensis* and several samples of *sapiens* mitochondrial DNA, it is clear that the two are similar: of 360 bases, 335 are identical and only 25 are different. But this is not really the surprise. When the same regions of DNA are compared from humans all over the globe, the DNAs differ by no more than 8 bases. All human DNA differs by less than 0.1%. By comparison, humans and chimpanzees differ by 55 bases. What this means is that the evolutionary distance between *H. neanderthalensis* and *H. sapiens* is at least three times the distance that separates the most different modern humans, and half the distance that separates humans from chimps. In other words, Neanderthals were very different from us. Furthermore, the similarity between Neanderthal and European DNA is no greater than the similarity between Neanderthal and Asian, Native American, Australian, or Oceanic DNA, arguing against the hypothesis that Neanderthals interbred with Europeans. The conclusion drawn from this evidence is that Neanderthals did not frequently interbreed with modern humans. Furthermore, the last ancestor shared by Neanderthals and modern humans existed over 200,000 years ago. More recently, the DNA of a second Neanderthal, found in the Caucasus Mountains in Russia, was analyzed. This sample was extremely similar to that of the first Neanderthal, notwithstanding a distance of approximately 1000 miles and a difference in age of as much as 70,000 years; and it, likewise, was not similar to that of that of modern humans. More recent analysis of Y-chromosome DNA has confirmed these initial findings. There are traces of Neanderthal-sequence genes in modern humans, particularly in Europeans (where the Neanderthals lived), so that if those traces do not go back to the common ancestor, then there was some interbreeding, but either it was very infrequent or the children did not do well.

Thus the Neanderthals, after a reign on earth approximately three times that of modern humans, disappeared. With, like us, the physiology of a tropical animal, they were capable of withstanding the rigors of an ice age in Europe. They had considerable skills and it is not yet excluded that they had conceptual thought and buried their dead (if they did not simply discard them in a common location). Were they driven further, into less hospitable territory, by the more capable, skilled modern humans? Did modern humans bring in diseases that they could not resist? Could they speak? Did they have a religion? We do not know, but the book is still open. What we do know is that they were sufficiently different from modern humans that there apparently was almost no effort to interbreed, or no success at it.

Concerning the migration of modern humans, certainly we can impute the sequence of events from the appearance of peoples before modern migrations began to mix the races once again. Likewise, we can look for common features to suggest the appearance of the earliest modern humans. For instance, chimpanzees and most groups of humans have straight black hair and moderately pigmented skin. Thus deviations from these patterns are probably more recent innovations. Humans, but not apes, have light colored soles and palms, perhaps used like the tails of white- tailed deer as markers by which adults could signal children or be seen by children as they walk away or as signs as humans communicated with each other by hand signals. Finally, again by DNA analysis, headlice seem to have differentiated from pubic lice approximately 70,000 years ago, suggesting that these humans became sufficiently hairless at that time to make the migration of lice between the two sites a difficult excursion.

Again, DNA analysis both adds precision and adds a surprising twist. First, there is much greater diversity of DNA in Africa than elsewhere, dating back 150,000 years, suggesting that, as the fossil record suggests, modern humans arose there. (Typically, if you have a very mixed population in one location and one small group escapes that location and starts a new colony, the new colony will have only a small portion of all the variants that existed in the home population. All human DNA shows similarity to a presumed ancestral DNA, especially in the DNAs that are normally passed on without modification, the mitochondrial DNA passed on from mother to child and the Y chromosome DNA passed from father to son.) Somewhere between 150,000 and 100,000 years ago lived a mitochondrial Eve and a Y chromosome Adam, who gave rise to everyone we now consider to be human. However, they lived in Africa, not the Garden of Eden (Iraq), and—despite the conceits of European artists (Fig. 11.9) more likely looked like the Khoi-San of southwestern Africa. Note that this does not necessarily mean that there was only one Adam and one Eve, but that only one or very few of the original chromosomes have survived to this day.

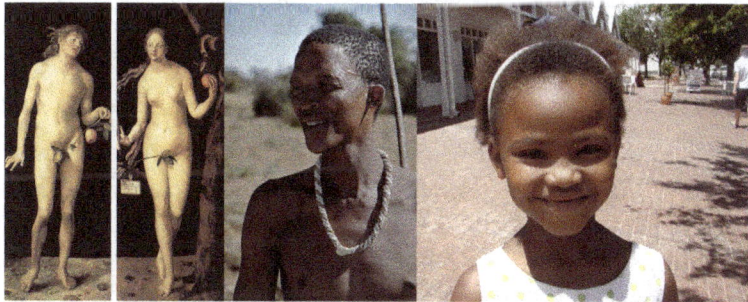

Figure 11.9. Contrary to the conceits of European artists (left: Adam and Eve as drawn by Albrecht Durer in 1504) the original humans almost certainly did not look like Europeans. They may have looked more like this [Khoisan Bushman](#)[viii] from South Africa or the Khoisan girl, as the Khoisan are one of the groups, all in Africa, who seem to have the largest number of ancestral genes. Judging from the apes, it may be more likely that the original human hair was straight.[ix]

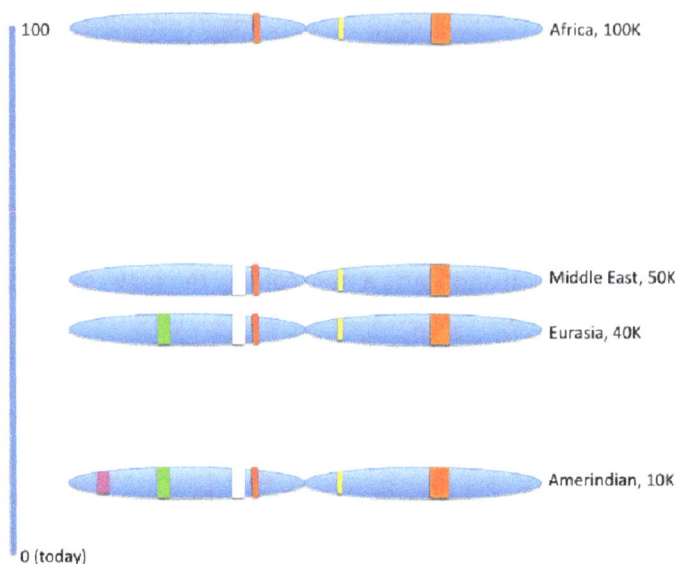

Figure 11.10. Human lineage as traced by Y chromosome markers. All human men, including those in Africa, Eurasia, and the Americas, share a pattern detectable by the use of specific stains (red, yellow, and orange markings). These markings therefore are ancestral. The pattern has been stable for approximately 100,000 years. New markers have arisen among the migrants, indicating their paths. One new mutation appeared 50,000 years ago (white) and is found in all men whose origin is outside of Africa before the beginning of slave trading. Therefore the migrants from Africa carried this trait, and the ancestral form is African. Another mutation (green) appeared about 40,000 years ago and is common among Eurasians, especially those from the Middle East through Central Asia, marking this migratory route. The magenta-colored mutation is found only in natives of the Americas. It appeared about 10,000 years ago, after this population had become established in the New World.

The DNA of humans outside of Africa comes from a subset of these types, arguing that a small band of humans left Africa 50-70,000 years ago and spread through Asia, Europe, and Australia by 30-40,000 years ago. Again, the sequence of mutations confirms the migratory patterns suggested by the fossil record and physical similarities (Fig. 11.10).

These sequences have helped us to understand human evolution and underscore some other points as well. The genetic similarity of all humans is remarkable and completely consistent with the history: We are essentially milli-cousins, or second cousins 1000 times removed. We are approximately 2000 generations separated from the first migration from Africa. In fact, there is less difference between any two humans than between chimpanzees from the East Coast and the West Coast of Africa, and more variation within a presumed "racial" group than between one group and another. All of this reflects on various attempts to display human evolution as a linear (hierarchical), as opposed to branched, process (see Chapter 5). Second, the wide variation in human appearance argues for evolution in small groups at relatively high selection pressure[x]. This would be consistent with our impression that humans have always lived in relatively small tribes.

Of the order of 5,000 years ago, a major change occurred in human style. Humans in the Middle East and in Asia learned to control the breeding of certain plants and animals, thus achieving domestication. Shortly thereafter, MesoAmerican and South American peoples marched down the same road. Domestication of plants tied populations to the agricultural fields and provided potentially stable sources of food, as chronicled in the story of Joseph and the Pharaoh (Genesis 39-42) and similar stories in other cultures, while domestication of animals provided sources of milk and meat, as well as the power to undertake large-scale building projects and the speed to move rapidly over great distances. As is described by Jared Diamond, these achievements made possible the rise of cities and perhaps led to selection that made these groups so successful.

Date (years ago)	Event	% of Age of Earth
-6,000,000,000	Origin of Universe	133
-4,500,000,000	Origin of Earth	100
-3,500,000,000	Origin of life	77
-1,000,000,000	Multicellular fossils	22
-600,000,000	Origin of animals	13
-500,000,000	Cambrian explosion	11
-350,000,000	Land vertebrates	8
-150,000,000	Origin of mammals; age of dinosaurs	3
-65,000,000	End of dinosaurs	1.4
-5,000,000	Humans, chimps emerge	0.1
-1,600,000	Genus Homo	0.04
-200,000	Genetic evidence for origin of Homo sapiens	0.004
-70,000	H. sapiens, jewelry	0.0015
-30,000	First art	0.0007
-15,000	Humans reach new world	0.0003
-10,000	Domestication of cattle	0.0002
-5,000	First cities	0.0001
-3,000	First writing	.000007
-150	First understanding of evolution	0.000003

Figure 11.11 Relative scale of time in evolutionary history. Humans are a very tiny fraction of total evolutionary history. Humans have been on this planet for an infinitesimal amount of time since the beginning of the earth. In that time, we have been able to write about our history for a period of less than 3% of our existence, and for 99.9% of that time we have had very little to no idea of how it came about The comment "Genus Homo" refers to the appearance of the large-brained modern versions, H. neanderthalensis rather than the earliest, much smaller-brained H. ergaster, who appeared approximately 1.6 million years ago

Think of what has happened to take us to this stage (Fig. 11.11). If the entire history of the earth were encompassed in the space of one day, the length of time that we have had any understanding of the process would take place in less than 3 milliseconds, less time than the sound of a single clap of the hands.[xi]

Chapter 12: When did humans acquire a soul?

There are many religions in the world, a few more widely accepted than the others, and each has its own version of the origin of the earth, the heavens, and of humans. For some believers, what scientists describe as their belief poses no problem, as either the religious description is considered to be allegorical or the hypotheses are considered to be in different realms. For some religions, however, notably fundamentalist Protestant Christianity and conservative elements of Catholicism, Judaism, Islam, Hinduism, and other faiths, the scientific view and the religious view are held to be incompatible, and the scientific view even heretical. What are we to make of this?

There are basically three issues to address: **the realm of what science does, the definition of the word "theory," and, most fundamentally, the question of what we choose to accept to structure our world.**

WHAT SCIENCE DOES

Scientists observe, experiment, and analyze the mechanics of how things work. Philosophers and theologians ask why. We can tell you the mechanics of how a clock works and what uses it might be put to, but there are societies for whom the day is simply the period between sunrise and sunset, and the seasons change little if at all. In our sense, they have little concept of time: they eat when they feel hungry, they do not count birthdays, and work (building a house or a boat) is done when it is done. They do not plan for the future. They do not have words such as "want"; if they want, they take. If individuals from these islands were to ask a scientist why one would bother to fractionate a day, the scientist would not be the appropriate person to answer. A philosopher would do better. Similarly, most scientists would say, "I can tell you how I understand the evidence to indicate the mechanics of how we came to have so many species of animals and plants on earth today. If you say that God willed all this to happen, this is fine. I cannot test the hypothesis that God willed evolution to happen. Therefore it is not a subject to which I can respond. I can discuss with you how I think He did it."

As the philosopher-physician Maimonides said, "A miracle is not something that could never happen without the intervention of God. A miracle is something that could always have happened, but did not until God chose to make it happen." At this point we must accept the fact that there are differences in detail. The Judaic versions of Genesis (there are two, Genesis I, 1-27 and Genesis II, 15-23, which differ in the order of creation)[i], accepted in most details by Christians and Muslims, describe Creation in seven days, with what we would call evolution occurring in the last four days. Most scientists would argue that the last four days of creation were in fact the better part of one billion years.

Other than the assumption that one version is allegorical, misleading, or misinterpreted, this disagreement is not easily resolved, and it is inappropriate to attempt to persuade a reader to accept one or the other interpretation. The attitude, however, represents a fundamental difference in how one looks at life, and what one accepts as truth. We all accept as comprehensible what is familiar to us as children, and understand the workings of newer inventions and changes in society in terms of what we previously understood. The same is true for what we accept as the most solid basis for evidence. For humans, sight is the most important sense. If you hear an animal's call in the night, you may not know whether the sound is that of an insect, frog, or bird. However, sight of the creature making the sound will allow you to make what you consider to be certain identification. For a dog, scent takes precedence over sight. When I was a child, my dog would wait for me after school. He would see me from a distance and look inquisitively. If I did not make a sign of recognition, he would wait until I came closer, then come up and sniff me. Only then would he be convinced that I was home. We also know, from optical illusions, that sight can be easily tricked, and juries must frequently contend with witnesses' differing versions of the same incident. In these cases we can force our intellects to take precedence over our senses, but it is not easy. These allegories bring us to our major point here, the growing strength of rationalism in 17th and 18th C. Europe and North America. In most religions during a major part of their histories, there is only one truth, and this is presented through a gospel or other work of divine origin. However, its meaning is sometimes ambiguous, and the change of society—for instance, increasing urbanization—may render agrarian images difficult to understand or even reveal errors. A priesthood is called upon to interpret the ambiguities.

The concept of the infallibility of prior highly respected sources may even extend to secular documents. For instance, the 2nd Century Hellenistic physician Galen provided a guide for medicine that applied for 1200 years. Even if later anatomists identified obvious errors, the supremacy of Galen was never in doubt. The rationalization was as follows: Most cadavers offered for dissection were those of condemned criminals. As criminals, they were by definition deformed. Surely the anatomy would be as Galen said if one did a dissection of an upstanding member of society.

The challenge to this argument came from many sectors. The great 16th C anatomist, Andreas Vesalius, argued for studying what one saw and mocked following the teachings of Galen in a spectacular fashion that must have deeply offended most of his colleagues (Fig. 12.1). William Harvey, in 16th C England, used a combination of experimentation and deductive reasoning to demonstrate the true function of arteries and veins, thus clearly illustrating errors by Galen.

Figure 12.1. Vesalius' insult of his colleagues. Whereas normally the professor gave the lecture while a preceptor performed the messy and smelly work of dissection, in the frontispiece of Vesalius' Anatomy he is illustrated as performing the dissection, while the professor is mocked by having a skeleton placed in his position. In this manner Vesalius not so subtly argued that one must respect facts and observations rather than ordained wisdom handed down from the ancients

Two other very important elements were the growing strength of Protestantism and the continuing exploration of the earth. Protestantism, in addition to arguing that each individual could interpret the Scriptures without the intervention of a priesthood, was now an established alternative, and presented several somewhat different interpretations of the meaning of the Scriptures. Explorers, who included interpreters, linguists, and priests, brought back stories of societies with very different stories of creation. The physicists who helped them plan their voyages, by analyzing the nature of motion and force (so that the sailors could design and use sails more effectively) and the motion of the planets (to help them find their way home) were applying these same findings to our planet and beginning to question whether the sun could really stand still, as in Joshua, and whether the earth was truly the center of the universe.

The rise of Protestantism as a religious and political, hence military, force meant that it was no longer possible for states to be the standard-bearers of a single religion. If a state was to govern a large population and not be perpetually at war, it would be obliged minimally to tolerate the existence of followers of a different religion. This movement culminated in the resurrection in Europe, after 1400 years, of a vision of a state as secular, an idea legitimized by the American Constitution and the French Revolution. Competing interpretations of sacred works were available for comparison. The many sources of alternative interpretations meant that it became possible to suggest that a higher standard for truth was analysis and logic. For the first time, people were suggesting that if logic and analysis contradicted Scriptures, it was possible that Scriptures were wrong.

This remains a fundamental difference. Scientists, by virtue of what they do, give precedence to logic and analysis in interpreting the mechanics of how the world works. Fundamentalists of many faiths insist that the exact description of how the world works is given in a holy work, received, as it were, as an email attachment from God, perfect and unalterable, even in translation. Thus if Joshua (Joshua 10: 13-14) says that the sun stopped in the heavens, it did; this is not a figurative statement.[ii]

The final issue is the meaning of "theory". In common parlance, "theory" may mean little more than "guess": "Why is the sky blue?" "My theory is that the air is colored blue." This latter statement is identical to "My guess is that the air is colored blue" (and is wrong). To a scientist, "theory" has a much more restricted meaning. "Theory" defines an upper level of a highly structured series of levels of certainty, as defined by Popper. Fundamentalists and rationalists clash over this definition, as in the sense of "Evolution is only a theory." Fundamentalists interpret the word in the popular sense of "guess", while rationalists use it to imply a level of certainty near that of, for instance, being able to calculate the hypotenuse of a right triangle if one knows the length of the other two sides.

~~~~~

Chapter 13: The impact of evolutionary theory:

THE EUGENICS SOCIETY AND THE I.Q. TEST

Racism certainly did not arise as a consequence of evolutionary theory. Humans have, as far as we know, always feared or disdained other groups that they have encountered, and usually considered the new groups to be vastly inferior to themselves. We have only to look at our vocabularies: barbaric or barbarian (like the Berbers, the peoples of northwest Africa); vandals (an invading tribe from the East), thugs (a warrior group in India); muggers (another group in India); savages (wild or uncivilized people). Our movies and entertainment rarely describe encounters with people from space as pleasant or agreeable, with peace-loving creatures eager to share their riches. Therefore it should come as no surprise that, long before 1859, writers and scholars were assembling hierarchical lists of humanity, based more on an Aristotelian linear tree of life than on a Linnaeus-style branching tree. Of course, with Europeans doing the analyzing and ranking, Europeans would naturally rank at the top of the list. Among Europeans, obviously, such issues were argued as to whether the rather long-faced northern (Nordic, Slavic, Teutonic) peoples were superior to the rather round-faced (Celtic, Alpine, Mediterranean) peoples. Numerous efforts were undertaken to quantify these alleged superiorities and inferiorities. Such efforts included phrenology (reading the bumps on the skull, which were alleged to indicate certain aspects of character); measurement of total brain size (by filling skulls with shot—small lead pellets—or seeds and weighing the contents); relative size of various parts of the brain; measurement of relative length of arms or legs, or degree of prognathism; and even inferring that shape of hair, color of skin, shape and density of eyebrows, or overall hirsuteness were direct indicators of other features of human worthiness. Drawings were unashamedly distorted to emphasize the more "ape-like" features of the disfavored race. No matter that no ape has the tightly-curled hair of the African and Mediterranean groups: this was obviously a primitive or inferior feature. (Apes have straight hair, shorter but otherwise more like the hair of most of non-African humanity.) Behavioral characteristics derived from being on the unfortunate end of a master-slave relationship, such as servility or stoicism in the face of punishment, were read as innate characteristics. The most well-known of the assumptions of hierarchy was the invention and use of the now-discredited term "Mongolian Idiot" to describe individuals bearing an extra chromosome and afflicted with the abnormality now known as Down's Syndrome. Among the characteristics of this affliction is a deformation of the skin fold of the eyelid, giving a characteristic to the eye quite unlike that of an Asian eye but, to a western eye, similar enough to confirm a presumption. The presumption was that Asians ranked well below Occidentals in the level of development of humanity. They were servile, obsequious, and intellectually far inferior to Occidentals (their great civilizations were known but ignored) and derived from a more primitive state of humanity. Thus the fact that people with Down's Syndrome were typically of low intelligence merely confirmed their relationship to Asians—they were throwbacks to this earlier stage of low intelligence. Another, even more startling, example is to consider why the lightly-pigmented races of humankind should be called "Caucasian," or people deriving a mountain range in southwest Russia. The people who more-or-less belong to this category were first identified as living in a territory stretching from northern India through the Middle East, Africa north of the Sahara, and throughout Europe. J. F. Blumenbach, the father of Anthropology, first used the term in 1795. In 1758 Linnaeus had classified humans into four races: Americans (Native Americans), Europeans, Asians, and Africans. Granted, he described these classifications according to the prejudices of the day. Americans were red, choleric, and upright, ruled by habit; Europeans were white, sanguine, and muscular, ruled by custom; Asians were yellow, melancholy, and stiff, ruled by beliefs; and Africans were black, phlegmatic, and relaxed, ruled by caprice. Who wouldn't prefer the European?

However, Linnaeus did not believe in a strictly hierarchical structure and, in spite of the overtones of his descriptions, did not truly rank humans. Toward the end of the century, though, the general social attitude had changed. Europeans were exploring and exploiting the world, bringing greater and greater riches home; philosophers had pushed the ideas of individual worth, individual rights, and freedom, and these ideas had achieved fruition in the appearance of new and exciting governments in the United States and in France; the motions of the planets and many of the laws of physics were known and exploited to improve human welfare. Likewise chemists were learning to extract riches from the earth and turn them into other valuable products. The wonders of electricity and magnetism were beginning to be known, as were the properties of the air (experiments to understand vacuums and to identify the life-giving functions of oxygen were underway). The attitude among the intellectuals of Enlightenment Europe was of steady, rapid, linear, and inexorable progress. Thus Blumenbach, otherwise an ardent disciple of Linnaeus, was a bit uncomfortable. A four-pronged lineage of humanity could not readily be defined as a hierarchy of progress. Pondering this problem, he realized that there was, in fact, a fifth race of humans, which he called Malay. Now it was clear that there was a symmetrical, elegant hierarchy or pyramid. At the lower level, of course were the Oriental and African lines. Above these were two slightly higher lines, the Americans and the Malays. Finally, the pinnacle, the crowning masterpiece of all humanity, was (naturally) the Caucasian group. Truly, this pattern displayed the progress of the species *Homo sapiens*. (We don't really have to go above this level to the future, more perfect stage because through some mysterious process we have achieved perfection and need rise no higher.)

Why Caucasian? Because to Blumenbach the races radiated from a center of origin and (he acknowledged racial mixing) at the center of origin would be the purest and most beautiful representatives of the race. To quote his description of a skull from a woman who had lived in Georgia (now the Republic of Georgia, where the Caucasus Mountains are found):

> "...really the most beautiful form of skull which...always of itself attracts every eye, however little observant.... In the first place, that stock displays...the most beautiful form of the skull, from which, as a mean and primeval type, the others diverge by most easy gradations...Besides, it is white in color, which [the skull!—au] we may fairly assume to have been the primitive color of mankind, since...it is very easy for that to degenerate into brown, but very much more difficult for dark to become white."[i]

The late 18th and early 19th Centuries are replete with such writings, in which bald assumptions about the native superiority of Europeans or subgroups of Europeans were defended by mellifluous but circular or otherwise illogical arguments or evidence. Today the hoops and contortions that these writers went through to make their point would be hilarious were they not so painful. Such arguments not only assuaged and flattered European egos, they served very effectively to convince Europeans of the justness and appropriateness of their treatment of natives of other lands. Servility, enslavement, and even slaughter were not crimes against humanity since the subject peoples could really aspire to no higher status and could not meaningfully exploit the resources of their lands. In fact, slavery might even be considered to be a type of beneficence, removing the slaves from the savagery and squalor from which they came. Tribal conflicts were obvious evidence of savagery whereas European wars were noble and justified. Such arguments and rationalizations readily justified the European conquests and exploitation of foreign lands and peoples but, with the continued faith in progress and the newly-arising sense of evolution, they came to the support of another fear, miscegenation (mixing of races). There had always been some concern, though not enough, of course, to prevent sailors, explorers, and soldiers from partaking of the pleasures of sexual encounters with women of different races. From the dawn of recorded history and, following the genetics, as far back in human history as we can trace, part of the spoils of war has been the women of the conquered people. However, where economics and social imperatives determined, the children of such encounters were routinely assigned to the subservient, slave or dependent, group. Even so, if miscegenation was not a desirable state of affairs, the problem was essentially the result of the licentiousness of the foreign women. Sir Thomas Browne, a physician and philosopher in the mid- 17th C, had debunked such myths as those that maintained that beavers castrated themselves to avoid capture; that the legs of badgers were shorter on one side than the other; and that, because Eve was created from the rib of Adam, men had one fewer rib than women. (This latter argument he addressed on two levels: first, by simply counting ribs on skeletons; and, second, by raising the question of how an excision from one man would be passed to his children, since someone who had lost one eye would still have a two-eyed child and those who had lost a limb would still have normal children. In spite of the clarity of his observation, the argument persisted 250 years later) He also debunked the myth that Jews had a peculiar odor, using several interesting arguments: You could not consistently detect an odor; there was no odor in synagogues; no one accused Jews who had converted to Christianity of having an odor; with intermarriage, the children did not have the alleged odor; and with intermarriage, you could not readily define a [genetic] population that you could call Jewish. Yet, even with this coldly analytical style, the image of miscegenation bothered him enough that he assumed that mixing of Jewish and Christian blood routinely arose from the desire of Jewish women to enjoy the pleasures of the far more desirable Christian men. (This sentiment was reciprocated. The Yiddish term "shiksa" arose as an intentionally derogatory term to describe a gentile (non-Jewish) woman whose seductive allure would entrap innocent Jewish men.) Two hundred years later (in 1863) Louis Agassiz, an important figure who is described elsewhere, could find no better explanation for the appearance of mixed-race (white-black) children than that women of African origin were particularly seductive and eager for sexual intercourse.

To this suspicious, prejudiced, and fear-mongering mix the prospect of evolution brought a new and frightening possibility: If humankind was engaged in a race to perfection, and the front-runners were the light-skinned races of Europe, then interracial marriage (or, more bluntly, sex) threatened to dilute and degrade the obvious superior traits of the dominant population. Societies that permitted the admixture of these poorer qualities would fall behind the purer (Caucasian) societies and would be derelict in their responsibility to continue the progress of humanity. Biracial and multiracial children were not only be an inconvenience and an embarrassment, they would be like an infection or a bad apple in the barrel, ultimately undermining the whole society. This new fear came to occupy the minds of many. But before one could truly act on this fear, there were of course certain niceties such as laws against cruelty and laws protecting freedom. To contravene these laws, it would be necessary to demonstrate that there really was no cruelty, that the lower races would not really benefit from the full protection of the law and that, in fact, they would benefit from being guided into their proper places (servants, slaves, or other dependent positions) by the paternalistic but beneficent supervision of their superiors.

Joseph-Arthur, comte de Gobineau, was a 19th C author who championed the idea that great civilizations fell when their citizens became corrupt enough to mix sexually with the degenerate populations that they had conquered. For the European-American civilization to maintain its hegemony, it must avoid mixing with Asian and African peoples. Gobineau's book was translated in the U.S. in 1856 (note: 3 years before the publication of *Origin of the Species*), just before the Civil War, and his translator emphasized in the preface that the U.S., which already had "the Indian" and "the negro" and was now threatened by "the extensive immigration of the Chinese" was triply threatened. Two years before, the translator had co-authored a best-seller entitled "Types of Mankind". The question was certainly a hot-button issue in the U.S.

Binet

Alfred Binet is the tragic figure in this story. Working in France at the end of the 19th and the beginning of the 20th C in France, he attempted to solve a problem for the government. Though he explicitly and emphatically stated the limitations of his solution, he opened Pandora's Box. The problem that the government had was that some students did not learn well, and the government wished to provide special education and training for those students. There were potentially three sources of their limitations. Some might simply be of such low intelligence that they could not learn; some might be sufficiently intelligent but, as peasants, lack skills or experience that urban recruits might have, in which case it would be necessary to identify the skills and alter the training; or their lack of literacy might simply make it difficult to give the students new information and for the students to retain important ideas (since they could not write them down as notes). Binet had worked with the great neuroanatomist Paul Broca and had attempted to assess intelligence as a function of brain size. Unlike many of the pretenders of his day, Binet had concluded that with truly unbiased measurements he could establish no correlation between intelligence and brain size. Thus he could not assess the children by measuring their head size. He therefore set out to devise a test that would discriminate among the three possibilities. It would have to test problem-solving ability without depending on written instructions or on experiences available only to certain groups. If he succeeded, the schools could weed out the untrainable and modify training procedures to accommodate different levels of literacy, skills, and experience. His test could produce a score that was indicative of something. He tried to make the test relate to the age of the child by getting mean scores for children of various ages. He assessed the child's relative ability subtracting the child's chronological age from the score the child achieved. All in all, he did an admirable job. Shortly thereafter the German psychologist W. Stern recommended dividing the score by the chronological age. In other words a 10-year old child who scored at the mean for 12-year-olds would score 12/10 or 1.2, which for appearance was multiplied by 100 to produce a score of 120.

Binet warned that the test identified simply the skills of an individual in a specific situation; that the skills and therefore the perceived intelligence could improve; and that the results were not generalizable to entire groups. A single test was very limited in value; one would need numerous tests, testing many functions, over different times. The ONLY function of the test was, in his eyes, to find the best means of teaching a child, never to categorize or restrict a child. In fact, he wrote with considerable anguish and frustration that some teachers assumed that the score, now described as an Intelligence Quotient or I.Q., represented an absolute and immutable assessment of the child's current and future worth:

> "The scale, properly speaking, does not permit the measure of intelligence because intellectual qualities are not superposable, and therefore cannot be measured as linear surfaces are measured."

And, again, arguing against teachers who claim that a student with a low I.Q. can "never" succeed:

> "Never! What a momentous word. Some recent thinkers seem to have given their moral support to these deplorable verdicts by affirming that an individual's intelligence is a fixed quantity, a quantity that cannot be increased. We must protest and react against this brutal pessimism..."

Binet's writings were translated and published in both England and in the United States, but his meaning sank in the Atlantic and the English Channel.

Goddard

Henry Herbert Goddard, of good family, finally brought Binet's test to the U.S., in the process as Stephan Jay Gould notes reifying the IQ score (turning the score into a single object of defined validity, asserting that it is a true, meaningful assessment of ability). He also managed to establish an argument that Binet never claimed, that the I.Q. was hereditary. Goddard ran, and was director of research at, the Vineland Training School for Feeble-Minded Girls and Boys in New Jersey. He eagerly used Binet's tests to assess his wards and, following his general attitude, to determine who was sufficiently unskilled and unreliable that they should not be trusted in society. Note the inversion: Binet devised the tests to determine how best to help students. Goddard wished to use the tests to determine whom to institutionalize. He also considered that the most important determinant of social deviation was low intelligence. Thus if one were to commit people based on their low intelligence, one would thereby reduce thievery, murder, and immorality. Goddard fervently believed that most people naturally found the level at which they functioned. Thus manual laborers were, for instance, naturally inferior to students at Princeton, and it would be a mistake to pretend that they could rise higher: "How can there be such a thing as social equality with this wide range of mental capacity?" Furthermore, this level of inheritance was likely inherited as a single gene and easily traceable as Mendelian inheritance, much like the yellow or green color of a pea.

To be fair, Goddard did not really understand the function of polygenic inheritance, the concept of which was elaborated in the 1930's. "Polygenic inheritance" means that many genes influence a character, and it is very difficult to assign a particular effect to a single gene. This is what is meant when physicians or scientists say that a disease "runs in families": your having a relative with the disease increases your chances of having the disease, meaning that you carry a trait that encourages the disease, but many other factors finally determine whether or not you manifest the disease. It is easy enough to see how this works. Suppose that you have a gene that destines you to be tall. However, you also carry a gene that prevents you from absorbing or using a specific vitamin; or a gene that causes you to digest food poorly; or a gene that interferes with the proper deposition of bone; or a gene that forces an abnormally early puberty so that your growth is terminated before you reach your full potential. With any of these other genes acting, you will never see your predestined height. A potential Picasso in a society with no access to the world's great art may never know what to do, or a brilliant woman scientist in a society that does not permit women to leave the house may wither unseen. We can put this another way: I have a marvelous gene that will renew all my organs at age 88, giving me the potential for living another vigorous 88 years. Unfortunately I am a soldier and am killed in battle at age 25. Perhaps if Goddard had been less convinced of the direct hereditability of intelligence he might have been more gentle in his assessment.

However, hereditability it was and, to preserve the high level of American society, according to Goddard those of low intelligence should be prevented from contributing to future generations. "It is perfectly clear that no feeble-minded person should ever be allowed to marry or to become a parent" (written in 1914). They should be committed to institutions like his own.

Furthermore, it was most likely that the immigrants coming into the country at this time also harbored large numbers of "defectives". It would also be important to prevent them from coming into the country. He visited Ellis Island and was convinced (by observation) that many morons and feeble-minded were applying for entrance into the country. He therefore trained two women ("The people who are best at this work, and who I believe should do this work, are women. Women seem to have closer observation than men. It was quite impossible for others to see how these two young women could pick out the feeble-minded…") to look for potentially feeble-minded immigrants to whom he would administer the Binet test. The results were astounding. 89% of the Jews, 80% of the Hungarians, 79% of the Italians, and 87% of the Russians were feeble-minded.

The tests were really quite remarkable. The immigrants were pulled from the line, given a pencil, shown a design, and then when the design was removed asked to draw it from memory. In addition to their confusion and fear, many had never held a pencil before. They were asked to state sixty words in three minutes, which most probably could have done if they had understood what the point of the question was. They were asked for the date, notwithstanding the fact that the Russian and Hebrew calendars, to name two, were different from the Julian calendar, and that they had been on a boat for two or more weeks. Goddard was thrilled that through his work in two years he had upped the number of rejections of immigrants by almost six fold.

He continued assiduously to affirm the inheritability of intelligence and general citizenship. His argument came from his tracing of the lineages of some of the inhabitants of his institution. One was particularly revealing. One group of "paupers and ne'er-do-wells" descended from an affair between an otherwise well-respected young man and a barmaid of presumed low intelligence and morality (the young man's morality was not at issue). The young man later married a good Quaker woman. The children of his marriage all became upstanding citizens. The descendants of the affair produced a startling number of "degenerates". Goddard described them as the descendants of Martin Kallikak, the name being derived from the Greek words kalos (good) and kakos (bad). (Our words such as calisthenics, calligraphy, and cacophony derive from the same roots). The barmaid produced the Kakos line, and the Quaker produced the Kalos line. The conclusions derived from the lineages were impressive: the fruits of the initial affair produced a huge cost to society, in dealing with all the degenerates and immoral descendants. It would have been much more practical to forbid that reproduction (Fig. 13.1).

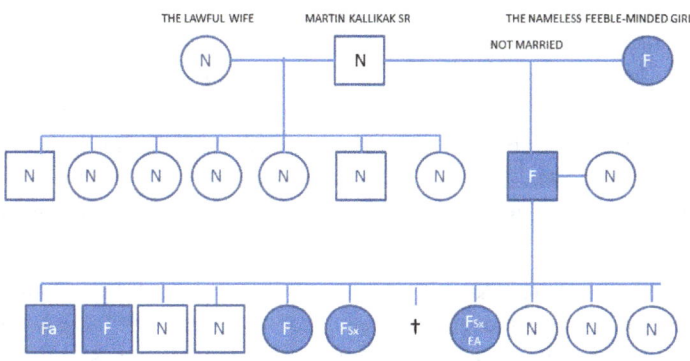

N: Normal F: Feeble minded Sx: Sexually Experienced A: Alcoholic

Figure 13.1. Kallikak lineage. A chart from Goddard's 1913 monograph. This graph follows the two lineages of Martin Kallikak Sr., from "the lawful wife" (left) and "the nameless feeble-minded girl" (right). Males are squares and females are circles. Other indications are listed below the chart, indicating that "the lawful wife" produced all healthy, productive descendants, while "the nameless feeble-minded girl" produced a large number of misfits. Another chart shows a later generation, indicating that one woman, though feeble-minded, produced only one feeble-minded son while the other woman produced a series of "defective" dependents. Credits: http://psychclassics.yorku.ca/Goddard/Developed by Christopher D. Green, York U

In his writing his fervor bespeaks a passionately concerned scientist, but in a terrifying way. He followed up the study of the Jukes family published in 1877 by Richard L. Dugsdale. In that paper, Dugsdale compared the genealogy of a disreputable family (the Jukes, a pseudonym) with that of an upright family (that of the stirring Quaker preacher Jonathan Edwards) and concluded that the tendency to be a social misfit was inherited. Goddard improved that dubious genetic analysis by comparing two lineages starting with the same father. From his illegitimate child, Goddard found 480 descendants in six generations. Most were a problem ([Fig. 13.2](#)).

His whole article reeks of preconceived notions and attitudes. He considers that most of the "Kakos" descendants who appear who appear to be normal are highly suspect and were probably misread. In any case, they have sometimes been transferred to other, "upstanding," families and this has been their salvation. (Goddard did not notice his obvious acceptance of environmental factors in this case.). The quotations are spectacular and speak for themselves:

> "This is the ghastly story of the descendants of Martin Kallikak Sr., from the nameless feeble-minded girl."

> "All of the legitimate children of Martin Sr. married into the best families in their state, the descendants of colonial governors, signers of the Declaration of Independence, soldiers and even the founders of a great university. Indeed, in this family and its collateral branches, we find nothing but good representative citizenship. There are doctors, lawyers, judges, educators, traders, landholders, in short, respectable citizens, men and women prominent in every phase of social life. They have scattered over the United States and are prominent in their communities wherever they have gone. Half a dozen towns in New Jersey are named from the families into which Martin's descendants have married. There have been no feeble-minded among them; no illegitimate children; no immoral women; only one man was sexually loose. There has been no epilepsy, no criminals, no keepers of houses of prostitution."

> "The foregoing charts and text tell a story as instructive as it is amazing. We have here a family of good English blood of the middle class, settling upon the original land purchased from the proprietors of the state in Colonial times, and throughout four generations maintaining a reputation for honor and respectability of which they are justly proud. Then a scion of this family, in an unguarded moment, steps aside from the paths of rectitude and with the help of a feeble-minded girl, starts a line of mental defectives that is truly appalling. After this mistake, he returns to the traditions of his family, marries a woman of his own quality, and through her carries on a line of respectability equal to that of his ancestors.

> "Clearly it was not environment that has made that good family. They made their environment; and their own good blood, with the good blood in the families into which they married, told.

> "Schools and colleges were not for them, rather a segregation which would have prevented them from falling into evil and from procreating their kind, so avoiding the transmitting of their defects and delinquencies to succeeding generations.

"At times she works hard in the field as a farm hand, so that it cannot be wondered at that her house is neglected and her children unkempt. Her philosophy of life is the philosophy of the animal.

"If all of the slum districts of our cities were removed to-morrow and model tenements built in their places, we would still have slums in a week's time, because we have these mentally defective people who can never be taught to live otherwise than as they have been living. Not until we take care of this class and see to it that their lives are guided by intelligent people, shall we remove these sores from our social life.

"There are Kallikak families all about us. They are multiplying at twice the rate of the general population, and not until we recognize this fact, and work on this basis, will we begin to solve these social problems.

"What can we do? For the low-grade idiot, the loathsome unfortunate that may be seen in our institutions, some have proposed the lethal chamber. But humanity is steadily tending away from the possibility of that method, and there is no probability that it will ever be practiced.

"But in view of such conditions as are shown in the defective side of the Kallikak family, we begin to realize that the idiot is not our greatest problem. He is indeed loathsome; he is somewhat difficult to take care of; nevertheless, he lives his life and is done. He does not continue the race with a line of children like himself. Because of his very low-grade condition, he never becomes a parent.

"It is the moron type that makes for us our great problem. And when we face the question, "What is to be done with them – with such people as make up a large proportion of the bad side of the Kallikak family?" we realize that we have a huge problem.

"The real sin of peopling the world with a race of defective degenerates who would probably commit his sin a thousand times over, was doubtless not perceived or realized. It is only after the lapse of six generations that we are able to look back, count up and see the havoc that was wrought by that one thoughtless act.

"When we conclude that had the nameless girl been segregated in an institution, this defective family would not have existed, we of course do not mean that one single act of precaution, in that case, would have solved the problem, but we mean that all such cases, male and female, must be taken care of, before their propagation will cease. The instant we grasp this thought, we realize that we are facing a problem that presents two great difficulties; in the first place the difficulty of knowing who are the feeble-minded people; and, secondly, the difficulty of taking care of them when they are known.

"A large proportion of those who are considered feeble-minded in this study are persons who would not be recognized as such by the untrained observer." [Note that here he takes the responsibility of making the judgment call himself—ral]

"In addition to this, the number would be reduced, in a single generation, from 300,000 (the estimated number in the United States) to 100,000, at least, – and probably even lower. (We have found the hereditary factor in 65 per cent of cases; while others place it as high as 80 per cent.)

"The other method proposed of solving the problem is to take away from these people the power of procreation. The earlier method proposed was unsexing, asexualization, as it is sometimes called, or the removing, from the male and female, the necessary organs for procreation. The operation in the female is that of ovariectomy and in the male of castration.

"There are two great practical difficulties in the way of carrying out this method on any large scale. The first is the strong opposition to this practice on the part of the public generally. It is regarded as mutilation of the human body and as such is opposed vigorously by many people. And while there is no rational basis for this, nevertheless we have, as practical reformers, to recognize the fact that the average man acts not upon reason, but upon sentiment and feeling;

"The question, then, comes right there. Should Martin Jr. have been sterilized! We would thus have saved five feeble-minded individuals and their horrible progeny.

"In considering the question of care, segregation through colonization seems in the present state of our knowledge to be the ideal and perfectly satisfactory method. Sterilization may be accepted as a makeshift, as a help to solve this problem because the conditions have become so intolerable. But this must at present be regarded only as a makeshift and temporary, for before it can be extensively practiced, a great deal must be learned about the effects of the operation and about the laws of human inheritance".

Category	Illegitimate line	Illegitimate line, further tracking	Legitimate line
Total	480	1146	496
Normal	46	197	496
Feeble-minded	143	262	0
Undetermined*	581	0	0
"Sexually immoral"		33	1
Alcoholic		24	2
Epileptic		3	0
Died in infancy		82	0
Criminal		3	0
"Kept houses of ill fame"		8	0

* "…we could not decide. They are people we can scarcely recognize as normal; frequently they are not what we could call good members of society."

Figure 13.2 Goddard's calculation of the fates of the descendants of the Kallikak family.

The number of quotes may be excessive, but they are important. Based on a fear of upsetting the wellbeing of the lives of the "good" people—Dugsdale had calculated that New York State had spent $1,308,000 [equivalent to over $26,000,000 today, as calculated from prices in 1913] between 1800 and 1875 on "social degenerates"— Goddard wanted to institutionalize at least 2% of the population and considered, really, that castration would be more effective. He was not a Nazi, and he was not crazy.

There were some problems with Goddard's data. Since he did not have access to many of the people, he would infer their morality "from the similarity of the language describing them to that used in describing persons she has seen". Others were "reputed to be a horse thief," "sexually amoral," or "wanton"— obviously social judgments bestowed on less-favored individuals, and not hard facts.

Even the facts were questionable. Others who have attempted to retrace the lineages note that many of the Kakos line did well for themselves and, in spite of everything, were upstanding citizens, but for some reason were not counted in the evaluations. The surviving photographs of the Kakos line show people who, to a modern eye, do not show the dullness and lack of interest that might betray someone of low intelligence (Fig. 13.3 left). Even worse, it is now reasonably certain that the pictures of the Kakos members who were not in institutions were retouched to make them appear more threatening and alien (Fig. 13.3 right). Nevertheless, Goddard's conviction of the hereditability of intelligence (or feeble-mindedness) and the cost to society of mental defect made him an avid defender of the rising Eugenics Society.

Figure 13.3. The institutionalized Deborah Kallikak (left), who in this and other photographs does not look unintelligent. Kallikak children (right), looks perturbing, but the eyes were redrawn. Credits: Kallikak Deborah - Kallikak family: Kallikaks_deborah2.jpg (53KB, MIME type: image/jpeg) This is a file from the Wikimedia Commons. An image from Henry H. Goddard's The Kallikak Family, 1912. AND http://psychclassics.yorku.ca/Goddard /Developed by Christopher D. Green, Kallikak Malinda – Kallikak family: This is a file from the Wikimedia Commons. An image from Henry H. Goddard's The Kallikak Family, 1912. http://psychclassics.yorku.ca/Goddard/ by Christopher D. Green

Termin

Lewis M. Termin was also totally convinced of the existence and value of an intelligence quotient, and he made a major improvement. He simplified the test so that it could be easily administered by less-trained individuals, and he aggressively marketed it as a tool for assessing children under the name of the Stanford-Binet scale (since he was a professor at Stanford). He prepared and marketed two scales: "…either may be administered in thirty minutes. They are simple in application, reliable, and immediately useful in classifying children in Grades 3 to 8 with respect to intellectual ability. Scoring is unusually simple." It was widely successful. Termin considered it to be only fair and reasonable, but even necessary, that society should assess its members and allocate professions and careers based on this assessment. Of course, the test questions, and their answers, might get some of us—certainly me, and perhaps you as well—into trouble.

> "An Indian who had come to town for the first time in his life saw a white man riding along the street. As the white man rode by, the Indian said—'The white man is lazy; he walks sitting down.' What was the white man riding on that caused the Indian to say, 'He walks sitting down?"

(Think out your answer before referring to the endnote.)[ii] In England, thanks to the efforts of the equally assiduous Cyril Burt, the British developed the "11 plus" exam, in which children were tested at age 11 and, on the basis of this test, allocated toward technical or vocational schools, or towards preparation for college. Before these tests were finally eliminated in the 1960's under the auspices of the Labor Party (roughly at that time equivalent to a party of union members and therefore keenly aware of the consequences of social ostracism) they carefully selected boys and girls on the basis of pure intellectual and immutable standards.

Or maybe not. One question the answer to which depended purely on problem solving and which of course would not have any basis in social experience was the following:

> "In the following list, which does not fit: dog, cat, car, motorcycle?"

Again, try to answer the question before looking at the endnote.[iii]

This sort of inanity continues steadily. There were efforts to assess the I.Q.'s of historical figures. When the assessments did not match the predictions, corrections were imposed so that the results came out "right". When looking at correlations with social status, intellectual deprivations in orphanages were considered trivial; schools were considered to be everywhere equal; greater familiarity with another language was not relevant. To his credit, Termin finally became so enmeshed in the entanglements of these rationalizations that, by the end of his life, he conceded that he was measuring environmental, not innate factors. But much damage had been done.

Robert M. Yerkes, a psychologist at Harvard deeply concerned that psychology did not get the respect that it deserved, argued for greater quantitation and emphasis on function. Consequently, addressing for the army the same issues that had been brought to Binet, persuaded the army to test 1.75 million men in World War I. The results were as feared: many recruits were morons, idiots, or feeble-minded. They could not answer questions of general knowledge, allegedly independent of culture, such as:

> Crisco is a: patent medicine, disinfectant, toothpaste, food product
>
> The number of a Kaffir's legs is: 2, 4, 6, 8.
>
> Christy Mathewson is famous as a: writer, artist, baseball player, comedian.

In looking at drawings to fill in what was missing, it was assumed that everyone would notice a missing rivet on a pocket knife or a filament in a light bulb. A drawing of a house might lack a chimney, but if a Sicilian answered by adding a crucifix, that would be wrong. And of course, an anxious, sometimes rural, often immigrant, army recruit would not get rattled by the following question and would correctly answer it within ten seconds:

> "Attention! Look at 4. When I say 'go' make a figure 1 in the space which is in the circle but not in the triangle or square, and also make a figure 2 in the space which is in the triangle and circle, but not in the square. Go."

The tests were analyzed many ways, including assessing the results by race and social background. The facts that many recruits scored 0; that the scores correlated heavily with geography; that recent immigrants did far more poorly; that Blacks from the North did far better than blacks from the South; and other indicators of heavy bias were not considered but rather rationalized out of existence. For instance, the differences between Northern and Southern Blacks were explained by the argument that only the brightest Blacks were smart enough to move to the North. The argument continued with a student of Yerkes, C.C. Brigham, who published a book, *A Study of American Intelligence*, in 1923. To him, the results were completely consistent with what one knew about the races. It was known that Jews had scored (in English, at Ellis Island) among the lowest of the immigrants, but by 1923 there were many Jews of undoubted abilities who were prominent in the U.S. The reason that they were prominent was because Jews were so routinely poor performers that the occasional success was truly startling, rather as we would notice a man who stood 5"7" but was surrounded by pygmies.

It might all sound laughable, but it led to two of the worst laws in American history as well as to the rise of Nazism in Europe. The country was already, following the First World War, fairly xenophobic, and hordes of immigrants were arriving, mostly from southern and eastern Europe. Francis Galton, a cousin of Darwin and an important player in the development of statistics, had argued for the hereditability of intelligence and described a philosophy he called eugenics, or good breeding. According to this philosophy, societies should strive to assure that only the fittest (note that he considered this word to be equivalent to "best") members procreate and leave young to the next generation. In this manner the human species will continuously improve. Carl Brigham,[iv] founder of the S.A.T., was a believer in eugenics and an active member in the fledgling Eugenics Society, which espoused the antipodal philosophy that the poorest members should not breed, or should be excluded from the country. Even though he, like Termin, ultimately realized that he had talked himself into a circle, his influence was considerable. The British Eugenics Society was formed in 1907. The American Eugenics Society was founded in 1922, and the Human Betterment Foundation, in 1928. Their founding members included lawyers, bankers, economists, and professors and chancellors of universities. The Eugenics Society, including the psychologists as well as prominent geneticists such as Thaddeus Hunt Morgan (the developer of *Drosophila* genetics) appeared before congress arguing for the protection of the country, resulting in the passage of the Johnson Immigration Restriction Act of 1924. This act reduced the then current immigration, which already had been somewhat curtailed. The law of 1921 allowed, per year and per country, an entry of immigrants equal to 3% of those already present from that country. The 1924 act dropped the number to 2% and used as its baseline the numbers present at 1890, thereby substantially reducing the numbers from southern and eastern Europe. As Calvin Coolidge commented, "America must be kept American". The laws were not revised until the mid-1960's, once again allowing immigration from southern and eastern Europe, as well as from Asia and Latin America. The expansion of the Asian immigration into the US dates from that change.

The other truly horrifying result was a collection of laws based on the assumption that inherited defects caused a huge financial drain on society and that it therefore was in society's interest to guarantee that these defects not be propagated. In other words, those deemed likely to generate cost for society should be sterilized. The issue finally reached the Supreme Court in 1927, with an issue that a young, apparently feeble-minded, woman had given birth to a likewise feeble-minded child. The woman's mother was also of dubious intelligence, and the state of Virginia requested the right to castrate her. In deciding for the state, the otherwise highly respected Oliver Wendell Holmes, Jr. wrote,

> "We have seen more than once that the public welfare may call upon the best citizens for their lives. It would be strange if it could not call upon those who already sap the strength of the state for these lesser sacrifices...Three generations of imbeciles are enough."

It really was almost like the right of cities to require the neutering of free-running dogs. It does not take much extrapolation to move from this attitude to the conclusion that whole races should be annihilated. We were not far from a Nazi attitude. The last known sterilization forced by these laws took place in the U.S. in the 1960's. Another horrifying aspect was that these policies were not promulgated by obvious kooks, white supremacists, or jingoists. Many prominent scientists, including leaders in psychology and some of the best biologists, argued for these laws. In more recent times prominent scientists, though not biologists, such as Arthur Jensen, Richard Herrnstein, and Charles Murray, have attempted to resuscitate these arguments. It matters little that intelligence—whatever it is— is so polygenic that its heritability is either nonexistent or extremely hard to measure, or that, according to the evolutionist Richard Lewontin, all the differences among races amount to 6.3% of the total variation seen among humans. There is a direct, if inadvertent and unintended, trail from Darwin to Hitler. What matters is not that we should not explore these ideas, because exploration, curiosity, and understanding are what make us human, but rather a more general admonition. As scientists we can easily overlook the social attitudes that create a sense of obviousness, as has been seen in the case of chimpanzee and gorilla social structures[v]. For scientists and non-scientists alike, we need to remember that information is not morality. Access to information does not make the scientist a seer, and a society has a right to judge the moral and social value of that information.[vi][vii]

~~~~~

CONCLUSIONS

Where does this leave us? In today's world, there are many sources of "truth," including several major, not precisely identical, stories of our origins based on the Old Testament, but we also have our insight based on our intellect, our ability to reason, and our ability to experiment and collect data. From these latter, we learn a very different story of our origin, and we learn that we are tied to other living creatures and that we are most likely not alone in the universe. Humankind has been descending from our ego trip for a very long time. We first appreciated that there were other tribes that claimed that they had the attention of gods other than our own. We began to understand that the earth was not the center of the universe; nor was the sun the center of the heavens. Many species have come and gone on our planet, including some very like us who, for all we know, may have been capable of aspirations that we would consider to be thought. Astronomers tell us that there may be 500,000,000 planets in the "Goldilocks zone" in the Milky Way, which itself is only one of millions of galaxies. We conclude that, even if we are unique, there are likely to be others of similar capabilities in the universe. It is harder and harder to convince ourselves that we, let alone a specific ethnic or religious subset of us, are the specific focus of a supreme being. Even if we are, I for one find it impossible to conceive of a god who would bestow curiosity, insight, and judgment on us and yet not expect us to use these powers to better our lot in the world we have. To attempt to do so—to be a scientist—is almost a religion to itself, and not to ask the questions we do would be impious to the extreme.

~~~~~

In this book I have tried to paint a picture of science, specifically biological science, as a product of its time and place, deriving from the social currents of the time and contributing to those same currents. Scientists generate ideas because they are curious, and they ask the questions that can be asked in the context of what is known and what is possible. Through these stories we can build some idea of how we got to where we are today, in an era in which the scientific way of thinking prevails but the extent to which it should be controlling is still hotly contested. We have not addressed the questions of how modern science is done: what are the criteria for evaluation; how these rules were developed and how they operate, for instance to trace the origin of AIDS and other diseases; how we improved our ability to measure things over one billion fold in the last fifty years; and where the tools and ideas of the now headline-making molecular biology arose, how they work, and where they are going. As in the case of the stories I tell here, most of the arguments that we can pursue to cover these subjects consists of anecdotes and stories of science. Please join me and continue the exploration of what makes a scientist in *Born This Way: How Science is Done and Practiced.*

###

About the author:

Richard A Lockshin was born in Ohio and received his undergraduate and graduate degrees from Harvard. He taught at the University of Rochester School of Medicine and Dentistry and later at St. John's University in New York, and is currently Professor Emeritus at St. John's. As a research scientist he is known for his studies of programmed cell death or apoptosis, now a major research topic, a field of which he is considered to be a founder. He has well over one hundred research publications, including several technical books in the field. He resides with his wife on Long Island, New York.

Discover other titles by Richard A Lockshin at Smashwords.com:

Born This Way: How Science is Done and Practiced, by Richard A. Lockshin, due Fall 2013

The Joy of Science: Springer, 2007, available through Amazon

Several other technical books on cell death and on aging, available through Amazon

Connect with Me Online:

mailto:rlockshin@gmail.com

Facebook: http://facebook.com/richard.lockshin

Smashwords: http:// [link to your author page] (to find the address for your author page, simply click on My Smashwords [that's it!])

My blog: http://sayingsofthepreachers.net

******

Endnotes

[i] The Joy of Science, by Richard A. Lockshin, Springer, 2007 ISBN-10: 140206098XISBN-13: 978-140206098

[ii] See *Born This Way: How Science is Done and Practiced,* by Richard A. Lockshin, due Fall 2013.

[iii] http://en.wikipedia.org/wiki/File:Basilisk_aldrovandi.jpg

[iv] http://en.wikipedia.org/wiki/File:Martigora_engraving.jpg

[v] http://en.wikipedia.org/wiki/File:Ljubljana_dragon.JPG

[vi] http://en.wikipedia.org/wiki/File:POL_wojew%C3%B3dztwo_zachodniopomorskie_COA.svg

[vii] http://people.wku.edu/charles.smith/wallace/S043.htm

[viii] http://www.csmonitor.com/Science/2012/0223/Why-Einstein-s-special-theory-%20of-relativity-is-probably-still-correct

[ix] http://www.catholiceducation.org/articles/science/sc0050.html

[x] http://en.wikipedia.org/wiki/Strasbourg_Cathedral#Astronomical_clock

[xi] Consider also these references:

Browne, J, 2002, Charles Darwin, The Power of Place, Alfred A. Knopf, New York, NY.

Cutler, Alan, 2004 (2003), The seashell on the mountaintop. Plume, the Penguin Group, New York.

Gould, Stephen Jay, 2002, The structure of evolutionary theory. Harvard University Press, Cambridge MA.

Sobel, Dava, 1999, Galileo's daughter, Penguin Books, New York.

http://www.victorianweb.org/science/cuvier.html (Georges Cuvier (1769–1832

Mann, Charles G., 2005, 1491, Alfred A. Knopf, New York.

Teresis, Dic, 2002, Lost discoveries. The ancient roots of modern science from the Babylonians to the Maya, Simon and Schuster, New York

Chapter 2

[i] http://www.nzetc.org/tm/scholarly/tei-DarJour-_N66499.html

[ii] http://www.nzetc.org/tm/scholarly/tei-DarJour-_N75469.html

[iii] http://www.flickr.com/photos/ral_bornthisway/8635749476/in/photostream

[iv] http://books.google.com/books?id=CTkLAAAAYAAJ&pg=PA348&lpg=PA348&dq=%22My+geological+examination+of+the+country+generally+created+a+good+deal+of+surprise+amongst+the+Chileno%22&source=bl&ots=Inet_oqASL&sig=rt6DlrEZZogZf25l1kh3xKdCSOk&hl=en&sa=X&ei=-jWBUeHYKoXV0gG3nIDwCQ&ved=0CDUQ6AEwAA#v=onepage&q=%22My%20geological%20examination%20of%20the%20country%20generally%20created%20a%20good%20deal%20of%20surprise%20amongst%20the%20Chileno%22&f=false

[v] http://www.geo.cornell.edu/geology/faculty/RWA/research/current_research/chile-m-88-earthquake-page/darwins-description-of-the-.html

[vi] http://www.unavco.org/community_science/science_highlights/2010/M8.8-Chile.html

[vii] http://www.brazilianfauna.com/sloth.php

[viii] http://eikaiwa-blog.blogspot.com/2010/10/pre-historic-florida-animals.html

[ix] http://www.worldatlas.com/webimage/countrys/samerica/galap.htm

[x] http://en.wikipedia.org/wiki/Cape_Verde

[xi] http://ccsbio.blogspot.com/2010/02/proboscis-prediction.html

[xii] http://www.everythingabout.net/articles/biology/animals/arthropods/insects/butterflies_and_moths/darwin_s_hawk_moth_full.jpg

[xiii] Darwin, Charles, Diary of the Voyage of the H.M.S. Beagle (edited from the MS by Nora Barlow, Cambridge, University Press, 1933; Kraus Reprint Co., New York, 1969, 440 pp.

Darwin, Charles, 2004, The voyage of the Beagle, Introduction by Catherine A. Henze, Barnes and Noble, New York (notebooks first published 1909).

Larson, Edward J. 2001. Evolution's workshop. God and science on the Galapagos Islands. Basic Books
(Perseus Books Group), New York.
http://www.wku.edu/"smithch/index1.htm (Essay on Wallace from Western Kentucky University).
http://www.clfs.umd.edu/emeritus/reveal/pbio/darwin/darwindex.html (Darwin-Wallace 1958 paper on
Evolution, from University of Maryland).

Chapter 3

[i] http://www.huffingtonpost.com/2011/01/21/fall-of-rome-climate-change_n_810419.html
[ii] http://bristleconemotel.webs.com/Right%20by%20Big%20Pine.jpg
[iii] http://wattsupwiththat.com/2010/03/17/medieval-warm-period-seen-in-western-usa-tree-ring-fire-scars/
[iv] http://www.flickr.com/photos/ral_bornthisway/8699486644/in/photostream
[v] http://www.ucmp.berkeley.edu/history/steno.html
[vi] http://en.wikipedia.org/wiki/File:Stenoshark.jpg
[vii] http://www.montauklighthouse.com/erosion.htm
[viii] http://disc.sci.gsfc.nasa.gov/geomorphology/GEO_5/geo_images_D-1/PlateD-1.jpeg
[ix] http://en.wikipedia.org/wiki/Niagara_river
[x] http://en.wikipedia.org/wiki/Mississippi_River_Delta
[xi] http://www.othemts.com/mdi/mountdesertisland51.jpg
[xii] http://www.flickr.com/photos/ral_bornthisway/8636177190/in/photostream
[xiii] http://zapatopi.net/kelvin/papers/on_the_secular_cooling_of_the_earth.html
[xiv] http://www.m4040.com/Knifemaking/Steel2.htm
[xv] http://www.chemistryviews.org/SpringboardWebApp/userfiles/chem/image/2011_October/Roth/fig1.jpg
[xvi] ttp://webbtelescope.org/webb_telescope/science_on_the_edge/cosmological_redshift.php
[xvii] http://www.youtube.com/watch?v=wrzWAox8NCM
[xviii] http://www.acs.psu.edu/drussell/Demos/doppler/doppler.html
[xix] http://www.youtube.com/watch?v=UI7VspSZeWw
[xx] http://galileoandeinstein.physics.virginia.edu/more_stuff/flashlets/doppler.htm
[xxi] http://galileoandeinstein.physics.virginia.edu/more_stuff/flashlets/doppler.htm
[xxii] http://www.rkm.com.au/animations/animation-Doppler-Effect-Star.htm
[xxiii] http://webbtelescope.org/webb_telescope/science_on_the_edge/cosmological_redshift.php
[xxiv] http://www.rkm.com.au/animations/animation-Doppler-Effect-Star.html
[xxv] http://pubs.usgs.gov/gip/dynamic/historical.html
[xxvi] http://www.flickr.com/photos/ral_bornthisway/8554667421/in/photostream
[xxvii] http://education.sdsc.edu/optiputer/flash/seafloorspread.htm
[xxviii] See *Born This Way: How Science is Done and Practiced,* by Richard A. Lockshin, due Fall 2013
[xxix] http://upload.wikimedia.org/wikipedia/commons/thumb/5/52/Pacific_Ring_of_Fire.svg/350px-Pacific_Ring_of_Fire.svg.png
[xxx] See *Born This Way: How Science is Done and Practiced,* by Richard A. Lockshin, due Fall 2013
[xxxi] Dalrymple, G. Brent, 1991, The Age of the Earth: Stanford, Calif., Stanford University Press, 474 p.
Darwin, C., The Origin of the Species
Harland, W.B.; Armstrong, R. L.; Cox, A.V.; Craig, L.E.; Smith, A.G.; Smith, D.G., 1990. A Geologic Time Scale, 1989 edition. Cambridge University Press: Cambridge, p.1–263
Powell, James Lawrence, 2001, Mysteries of Terra Firma: the Age and Evolution of the Earth, Simon & Schuster
Wilde S. A., Valley J.W., Peck W.H. and Graham C.M. (2001) Evidence from detrital zircons for the existence of continental crust and oceans on the Earth 4.4 Gyr ago.
Nature, v. 409, pp. 175–178 http://nationalacademies.org/evolution/ (free downloads) http://pubs.usgs.gov/gip/geotime/age.html
McPhee, John (1982), Basin and Range, Farrar, Straus and Giroux, New York; Reissue edition
McPhee, John (1984), In Suspect Terrain, Farrar, Straus and Giroux; Reissue edition (January 1, 1984)
http://pubs.usgs.gov/gip/dynamic/world_map.html (Explanation of continental drift from US Geological
Survey)

Chapter 5

[i] http://en.wikisource.org/wiki/An_Essay_on_the_Principle_of_Population

[ii] For these latter three, see See *Born This Way: How Science is Done and Practiced*, by Richard A. Lockshin, due Fall 2013

[iii] http://www.econlib.org/library/Malthus/malPlong

[iv] http://www.econlib.org/library/Malthus/malPop.html

[v] http://maps.grida.no/go/graphic/evolution_of_the_world_grain_production_comparison_world_europe_china_africa

[vi] http://earthtrends.wri.org/pdf_library/country_profiles/agr_cou_706.pdf

[vii] http://maps.grida.no/go/graphic/evolution_of_the_world_grain_production_comparison_world_europ

[viii] http://maps.grida.no/go/graphic/food_production_index

[ix] http://en.wikipedia.org/wiki/Criollo_horse

[x] http://www.google.com/publicdata/explore?ds=d5bncppjof8f9_&met_y=sp_pop_totl&idim=country:SO%20M&dl=en&hl=en&q=population+of+somalia#ctype=l&strail=false&bcs=d&nselm=h&met_y=sp_pop_totl

[xi] To Europeans, the contract was as in Genesis I, 26: "Then God said, 'Let us make man in our image, after our likeness; and let them have dominion over the fish of the sea, and over the birds of the air, and over the cattle, and over all the earth, and over every creeping thing that creeps upon the earth.'" Other cultures had different views but, to Europeans, all organisms served humans under God

[xii] http://upload.wikimedia.org/wikipedia/commons/f/fe/Mimosa_Pudica.gif

[xiii] http://en.wikipedia.org/wiki/File%3AVelvetwormonleaf.gif

[xiv] http://upload.wikimedia.org/wikipedia/commons/3/33/Northern_leopard_frog_1.jpg

[xv] http://en.wikipedia.org/wiki/File%3ARana_sphenocephala.jpg

[xvi] http://www.nationalgeographicstock.com/comp/MI/001/1235253.jpg

[xvii] http://allaboutfrogs.org/info/species/leopard.html

[xviii] Photo courtesy of Igor Siwanowicz, http://photo.net/photodb/user?user_id=1783374, reprinted with permission.

[xix] http://commons.wikimedia.org/wiki/File:Pterodactyl_%28PSF%29.png

[xx] http://en.wikipedia.org/wiki/File:Homology_vertebrates.svg

[xxi] http://en.wikipedia.org/wiki/File:Whale_skeleton.png

[xxii] http://embryo.soad.umich.edu/carnStages/stage16/Opticals/rtLat3.html

[xxiii] R. W. Hegner's 1909 Leptinotarsa The origin and early history of the germ cells in some chrysomelid beetles and Drosophila: https://play.google.com/books/reader?id=gHhNAAAAMAAJ&printsec=frontcover&output=reader&authuser=0&hl=en&pg=GBS.PA299

[xxiv] http://www.sciencedaily.com/releases/2009/08/090803083916.htm

[xxv] http://www.wellcome.ac.uk/Education-resources/Education-and-learning/animations/dna/wtdv026689.htm

[xxvi] Consider also these references:

http://nationalacademies.org/evolution/ (Essay on Evolution from the National Academy of Sciences)

http://darwiniana.org/(Darwiniana and Evolution, International Wildlife Museum, Tucson, AZ)

http://www.pbs.org/wgbh/evolution/index.html (Summary of Evolution series from Public Broadcasting System)

Mayr, E (1963) Animal Species and Evolution, Harvard University Press, Cambridge.

Diamond, J.M. (1965) Zoological classification system of a primitive people. Science 151: 1102–1104.

http://www.amnh.org/exhibitions/expeditions/treasure_fossil/Treasures/Giant_Sequoia/sequoia.html?acts (Site from American Museum of Natural History, New York, image of giant sequoia tree rings)

http://www.ac.wwu.edu/~stephan/malthus/malthus.0.html (Malthus' complete essay, from Western Washington University)

Browne, J., 2002, Charles Darwin, The power of place, Alfred A. Knopf, New York.

Darwin, Charles, 2004 (1859) The origin of the species, Introduction and notes by George Levine, Barnes and Noble Classics, New York.
Eldredge, Niles, 2005, Darwin, Discovering the Tree of Life, Norton and Company, New York.
Gould, Stephen Jay, 2002, The structure of evolutionary theory. Harvard University Press, Cambridge MA.
[ant lion] Zim H. S. and Cottam, C, (Irving, JG, Illustrator) Insects. A Guide to Familiar American Insects, Simon and Schuster, New York, 1956.

Chapter 6

[i] http://www.flickr.com/photos/ral_bornthisway/8510434631/sizes/k/in/photostream/
[ii] http://babel.hathitrust.org/cgi/pt?id=coo.31924090300249;view=1up;seq=509
[iii] http://www.linnean.org/contents/history/dwl_full.html
http://www.clfs.umd.edu/emeritus/reveal/PBIO/darwin/ (Darwin-Wallace 1858 paper, from Univ. of Maryland)
http://www.darwiniana.org (Essay on Darwin and Evolution, from International Wildlife Museum, Tucson, AZ)
http://nationalacademies.org/evolution/ (Essay on evolution, from National Academies of Science)

Chapter 7

[i] http://en.wikipedia.org/wiki/Hugo_de_Vries
[ii] http://en.wikipedia.org/wiki/Carl_Correns
[iii] http://en.wikipedia.org/wiki/Erich_von_Tschermak
[iv] See *Born This Way: How Science is Done and Practiced*, by Richard A. Lockshin, due Fall 2013.
[v] http://upload.wikimedia.org/wikipedia/commons/9/94/Preformation.GIF
[vi] See *Born This Way: How Science is Done and Practiced*, by Richard A. Lockshin, due Fall 2013.
[vii] http://www.flickr.com/photos/ral_bornthisway/8579831783/in/photostream
[viii] http://www.indiana.edu/~oso/lessons/Genetics/Phenotypes.html
[ix] http://www.mendelweb.org/ (Mendel's original paper, from an international resource for the web)

Chapter 8

[i] http://www.geocities.com/CapeCanaveral/Lab/2948/orgel.html
[ii] http://upload.wikimedia.org/wikipedia/commons/thumb/1/1b/Stromatolites_in_Sharkbay.jpg/300px-Stromatolites_in_Sharkbay.jpg
[iii] http://www.livescience.com/5355-big-blobs-change-view-evolution.html
[iv] http://www.flickr.com/photos/ral_bornthisway/8511619996/in/photostream
[v] http://www.flickr.com/photos/ral_bornthisway/8580972398/in/photostream
[vi] http://en.wikipedia.org/wiki/File:Black-band_ironstone_%28aka%29.jpg
[vii] http://www.geocities.com/CapeCanaveral/Lab/2948/orgel.html (Orgel's essay on Origin of Life)
http://www.ncseweb.org/icons/icon1millerurey.html (The Miller-Urey experiment, from National Center for Science Education)
Orgel, Leslie (1998) The origin of life. A review of facts and speculations. Trends in Biochemical Sciences 23: 491–495

Chapter 9

[i] MAP OF TECTITE DISTRIBUTION
[ii] http://www.dvidshub.net/image/684226/shaded-relief-with-height-color-and-landsat-yucatan-peninsula-mexico#.UYMhf5VxWfQ
[iii] http://palaeo.gly.bris.ac.uk/communication/hanks/eff.html
[iv] http://upload.wikimedia.org/wikipedia/commons/e/eb/Chicxulub-gravity-anomaly-m.png
[v] http://commons.wikimedia.org/wiki/File:K-T_boundary_at_Starkville_South.jpg

[vi] (http://upload.wikimedia.org/wikipedia/commons/7/70/Trinlake2.JPG).
[vii] http://upload.wikimedia.org/wikipedia/commons/5/5c/Two_tektites.JPG
[viii] http://upload.wikimedia.org/wikipedia/commons/thumb/5/52/820qtz.jpg/220px-820qtz.jpg
[ix] http://upload.wikimedia.org/wikipedia/commons/0/06/Extinction_intensity.svg
[x] Credits: Schmidt-Nielsen, K. Animal Physiology: Adaptation and Environment, Cambridge University Press, 1975;from P.F. Scholander, Biological Bulletin 1950, 99: 234

Chapter 10

[i] Photo courtesy of Igor Siwanowicz, http://photo.net/photodb/user?user_id=1783374, reprinted with permission.
[ii] Get reference from lab Firefox bookmark
[iii] http://www.fws.gov/jclarksalyer/early.htm
[iv] http://commons.wikimedia.org/wiki/File:Black_widow.svg
[v] See *Born This Way: How Science is Done and Practiced*, by Richard A. Lockshin, due Fall 2013.Lederberg
[vi] Homeotic genes are genes that contain a short sequence of bases that produces a short sequence of amino acids in proteins, allowing these proteins to bind to DNA. In binding, they establish positions along the axes of an animal: tip of the head to tip of the tail, from back to front, or from the center of the body to the periphery. They are present in all higher animals, though there is only one set in insects and several in mammals, and they perform the same functions. A gene that controls the formation of the head in a fruit fly does so as well in a mammal; and the genes from insects and mammals are so similar that they can sometimes be substituted for each other. See, for instance, http://en.wikipedia.org/wiki/Homeotic_gene
[vii] http://www.flickr.com/photos/ral_bornthisway/8704426171/in/photostream
[viii] http://www.macroevolution.net/stephen-jay-gould.html#.UNITwneKI18
[ix] http://www.transitionalfossils.com/
[x] http://aso.gov.au/titles/historical/tasmanian-tiger-footage/clip1/
[xi] Lizard limb credits

Chapter 11

[xii] Halder, G., Callaerts, P., and Gehring, W.J. (1995), New perspectives on eye evolution. Curr. Opin. Genet. Dev. 5: 602–609.
Halder, G., Callaerts, P., and Gehring, W.J. (1995) Induction of ectopic eyes by targeted expression of the eyeless gene in Drosophila, Science 267: 1788–1792.
Land, M. F. and Fernald, Russel D. (1992) The evolution of eyes. Ann. Rev. Neurosci. 15: 1–29.
Darwin, Charles, 2004 (1859) The origin of the species, Introduction and notes by George Levine, Barnes and Noble Classics, New York.
Gould, Stephen Jay, 2002, The structure of evolutionary theory. Harvard University Press, Cambridge MA.
Campbell, Neil A. and Reece, Jane B.(2004) Biology (7th Edition), Benjamin Cummings, Boston, MA
Gould, Stephen Jay, 2002, The structure of evolutionary theory. Harvard University Press, Cambridge MA.
Wilson, Edward O. 1995, Naturalist, Warner Books, New York.
Wilson, Edward O, 1996, In search of nature, Island Press/Shearwater Books, Washington, D.C.
Ameisen, Jean-Claude (2003), La Sculpture du vivant : Le suicide cellulaire ou la mort créatrice, Seuil, Paris (available only in French)
Diamond,Jared (1999, reprinted 2005) Guns, Germs, and Steel. The Fates of Human Societies, W.W. Norton, Co., New York.
Mann, Charles G., 2005, 1491, Alfred A. Knopf, New York.

Chapter 12

[i] See Wikipedia on Neanderthal, and Smithsonian (http://www.si.edu) on Shanidar cave (Neanderthal)
[ii] http://en.wikipedia.org/wiki/The_Great_Wave_off_Kanagawa
[iii] http://www.bousai.metro.tokyo.jp/english/e-knowledge/mechanism.html
[iv] http://en.wikipedia.org/wiki/Click_languages#Languages_with_clicks
[v] http://www.flickr.com/photos/ral_bornthisway/8738033981/in/photostream

[vi] For instance, potentially counting scratches and designs have been found on bone, handprints and dots in cave art, and various types of beads that presumptively are jewelry. Venuses (Figures 11.x and 11.xi) are among the several types of carving, mostly of animals.
[vii] http://hoopermuseum.earthsci.carleton.ca/neanderthal/n-skeleton.html
[viii] http://en.wikipedia.org/wiki/Khoisan

[x] Selection pressure is, roughly, the survival of one type compared to the survival of all types. For instance, if 99% of one variant survive to reproduce, compared to 100% of another variant, the selection pressure is 1%. If 90% survive, the selection pressure is 10%; if 10% survive, the selection pressure is 90% against. If the genetics is clean and the population types can be well controlled and measured, this number can be calculated from the distributions of the variants.
[xi] Further references
Shreeve, James, The Greatest Journey, National Geographic March 2006, pp 60–69.
http://www.cbsnews.com/stories/2005/03/18/60minutes/main681558.shtml?CMP=ILC-SearchStories (Sea Gypsies see signs in the waves)
http://www.sciencenews.org/articles/20030823/fob7.asp (The Naked Truth? Lice hint at a recent origin of clothing)
http://www.bonobo.org/whatisabonobo.html (description of bonobos, from an organization devoted to their preservation)
http://www.janegoodall.org/chimps/index.html (description of chimpanzees, by Jane Goodall)

http://www.culture.gouv.fr./culture/arcnat/chauvet/en/visite.htm (Chauvet caves (~32,000 years ago): In English, very nice)

http://www.cantabriainter.net/cantabria/lugares/cuevasaltamira.htm (Altamira (~15,000 years ago; In Spanish, but very good pictures, from the Cantabria Province of Spain).

http://vm.kemsu.ru/en/palaeolith/lascaux.html (Russian, French, and Spanish caves ~20,000–15,000 years ago)
http://witcombe.sbc.edu/willendorf/willendorfdiscovery.html (Venus of Willendorf, 23,000 years ago– site is very good, from Sweet Briar College)
http://www.hominids.com/donsmaps/galgenbergvenus.html (Venus of Brassempouy 28,000 BC?)
http://www.hominids.com/donsmaps/ukrainevenus.html (Berekhat Ram, Israel: –800,000 to – 233,000 years ago)

Chapter 12
[i] **Genesis 1**
The Beginning
1 In the beginning God created the heavens and the earth.
2 Now the earth was †a‡ formless and empty, darkness was over the surface of the deep, and the Spirit of God was hovering over the waters.
3 And God said, "Let there be light," and there was light.
4 God saw that the light was good, and He separated the light from the darkness.
5 God called the light "day," and the darkness he called "night." And there was evening, and there was morning—the first day.
6 And God said, "Let there be an expanse between the waters to separate water from water."
7 So God made the expanse and separated the water under the expanse from the water above it. And it was so.
8 God called the expanse "sky." And there was evening, and there was morning—the second day.
9 And God said, "Let the water under the sky be gathered to one place, and let dry ground appear." And it was so.
10 God called the dry ground "land," and the gathered waters he called "seas." And God saw that it was good.
11 Then God said, "Let the land produce vegetation: seed-bearing plants and trees on the land that bear fruit with seed in it, according to their various kinds." And it was so.
12 The land produced vegetation: plants bearing seed according to their kinds and trees bearing fruit with seed in it according to their kinds. And God saw that it was good.
13 And there was evening, and there was morning—the third day.

14 And God said, "Let there be lights in the expanse of the sky to separate the day from the night, and let them serve as signs to mark seasons and days and years,
15 and let them be lights in the expanse of the sky to give light on the earth." And it was so.
16 God made two great lights—the greater light to govern the day and the lesser light to govern the night. He also made the stars.
17 God set them in the expanse of the sky to give light on the earth,
18 to govern the day and the night, and to separate light from darkness. And God saw that it was good.
19 And there was evening, and there was morning—the fourth day.
20 And God said, "Let the water teem with living creatures, and let birds fly above the earth across the expanse of the sky."
21 So God created the great creatures of the sea and every living and moving thing with which the water teems, according to their kinds, and every winged bird according to its kind. And God saw that it was good.
22 God blessed them and said, "Be fruitful and increase in number and fill the water in the seas, and let the birds increase on the earth."
23 And there was evening, and there was morning—the fifth day.
24 And God said, "Let the land produce living creatures according to their kinds: livestock, creatures that move along the ground, and wild animals, each according to its kind." And it was so.
25 God made the wild animals according to their kinds, the livestock according to their kinds, and all the creatures that move along the ground according to their kinds. And God saw that it was good.
26 Then God said, "Let us make man in our image, in our likeness, and let them rule over the fish of the sea and the birds of the air, over the livestock, over all the earth, †b‡ and over all the creatures that move along the ground."
27 So God created man in his own image, in the image of God he created him; male and female he created them.

Genesis 2:
14 15 The LORD God took the man and put him in the Garden of Eden to work it and take care of it.
16 And the LORD God commanded the man, "You are free to eat from any tree in the garden;
17 but you must not eat from the tree of the knowledge of good and evil, for when you eat of it you will surely die."
18 The LORD God said, "It is not good for the man to be alone. I will make a helper suitable for him."
19 Now the LORD God had formed out of the ground all the beasts of the field and all the birds of the air. He brought them to the man to see what he would name them; and whatever the man called each living creature, that was its name.
20 So the man gave names to all the livestock, the birds of the air and all the beasts of the field. But for Adam !h" no suitable helper was found.
21 So the LORD God caused the man to fall into a deep sleep; and while he was sleeping, he took one of the man's ribs !i" and closed up the place with flesh.
22 Then the LORD God made a woman from the rib !j" he had taken out of the man, and he brought her to the man.
23 The man said, "This is now bone of my bones and flesh of my flesh; she shall be called 'woman, !k"' ' for she was taken out of man."

[ii] Further references:
Barbour, Ian G, 2000, When science meets religion, Harper, San Francisco.
Maimonides: Maimonides, Guide to the Perplexed (Chicago: University of Chicago Press, 1989), 2:25 For a miracle cannot prove that which is impossible; it is useful only as a confirmation of that which is possible,— http://www.sacred-texts.com/jud/gfp/gfp.htm Part 3, Chapter 24)
http://bibleontheweb.com/Bible.asp Genesis 1 (New International Version) New International Version (NIV) Copyright © 1973, 1978, 1984 by International Bible Society

Chapter 13

[i] Beside the remarkable assumptions implied here, to a biologist the final passage is wrong. Pale pigmentation generally indicates a failure to synthesize a pigment, as is described in Chapter 13, pages 186–187. It is much easier to lose a gene responsible for producing an enzyme to synthesize a pigment than it is to gain a gene to make the pigment. Albinos regularly appear among all human races, whereas mutations to darker skin are not seen. Furthermore, since all human races have pale

palms and soles, moving to lighter pigmentation requires merely expanding the regions where pigmentation is restricted

[ii] You of course recognized "bicycle" as the correct response. "Horse" is incorrect because an Indian would obviously recognize a horse; "wheel chair"; "on someone's back"; or any other clever or imaginative answer was wrong.

[iii] If my experience is any guide, welcome to vocational school. The correct answer is "cat" because it is the only one that does not require a license. The 11+ exam, incidentally, was instituted largely on the influence of Sir Cyril Burt, whose studies of the similarity of intelligence in identical twins was the primary basis for the belief that intelligence was highly hereditable. After his death in 1947, reexamination of his data led to the conclusion that almost all of it was fabricated.

[iv] http://en.wikipedia.org/wiki/Carl_Brigham

[v] See *Born This Way: How Science is Done and Practiced,* by Richard A. Lockshin, due Fall 2013

[vi] Barbour, Ian G, 2000, When science meets religion, Harper, San Francisco.
Maimonides: Maimonides, *Guide to the Perplexed* (Chicago: University of Chicago Press, 1989), 2:25
For a miracle cannot prove that which is impossible; it is useful only as a confirmation of that which is
possible,— http://www.sacred-texts.com/jud/gfp/gfp.htm Part 3, Chapter 24)
http://bibleontheweb.com/Bible.asp Genesis 1 (New International Version) New International Version
(NIV) Copyright © 1973, 1978, 1984 by International Bible Society
New York Times, Feb. 8 2003, Bad Seed or Bad Science: The Story of the Notorious Jukes Family
Available as http://www.wehaitians.com/bad%20seed%20or%20bad%20science.html
Goddard Henry Herbert (1913) The Kallikak Family: A Study in the Heredity of Feeble-Mindedness
Available as Classics in the History of Psychology An internet resource developed by Christopher D. Green, York University, Toronto, Ontario http://psychclassics.yorku.ca/Goddard/
http://en.wikipedia.org/wiki/Jukes_and_Kallikaks (Jukes and Kallikaks, from Wikipedia)
http://www.criminology.fsu.edu/crimtheory/week4.htm (A long, good essay on biological and social
factors in criminology, from Florida State University)
Estabrook Arthur H. 1916 The Jukes in 1915 Carnegie Institution of Washington (available as
http://www.disabilitymuseum.org/lib/docs/759.htm http://www.disabilitymuseum.org
Gould, Stephen Jay, 1981, 1996, The mismeasure of man, W.W. Norton and Co., New York, N.Y.
Eldredge, Niles, 2001 (2000) The triumph of evolution and the failure of creationism, Henry Holt &
Co., New York.

vii Further references: New York Times, Feb. 8 2003, Bad Seed or Bad Science: The Story of the Notorious Jukes Family Available as
http://www.wehaitians.com/bad%20seed%20or%20bad%20science.html; Goddard Henry Herbert (1913) The Kallikak Family: A Study in the Heredity of Feeble-Mindedness Available as Classics in the History of Psychology An internet resource developed by Christopher D.   Green, York University, Toronto, Ontario http://psychclassics.yorku.ca/Goddard/;
http://en.wikipedia.org/wiki/Jukes_and_Kallikaks (Jukes and Kallikaks, from Wikipedia);
http://www.criminology.fsu.edu/crimtheory/week4.htm (A long, good essay on biological and social factors in criminology, from Florida State University); Estabrook Arthur H. 1916 The Jukes in 1915 Carnegie Institution of Washington (available as http://www.disabilitymuseum.org/lib/docs/759.htm http://www.disabilitymuseum.org; Gould, Stephen Jay, 1981, 1996, The mismeasure of man, W.W. Norton and Co., New York, N.Y.; Eldredge, Niles, 2001 (2000) The triumph of evolution and the failure of creationism, Henry Holt & Co., New York.

www.ingramcontent.com/pod-product-compliance
Lightning Source LLC
Chambersburg PA
CBHW050551300426
44112CB00013B/1876